EAA Series

AF790072

EAA series is successor of the EAA Lecture Notes and supported by the European Actuarial Academy (EAA GmbH), founded on the 29 August, 2005 in Cologne (Germany) by the Actuarial Associations of Austria, Germany, the Netherlands and Switzerland. EAA offers actuarial education including examination, permanent education for certified actuaries and consulting on actuarial education.

actuarial-academy.com

More information about this series at http://www.springer.com/series/7879

Ermanno Pitacco

Health Insurance

Basic Actuarial Models

Springer

Ermanno Pitacco
University of Trieste
Trieste
Italy

ISSN 1869-6929 ISSN 1869-6937 (electronic)
EAA Series
ISBN 978-3-319-12234-2 ISBN 978-3-319-12235-9 (eBook)
DOI 10.1007/978-3-319-12235-9

Library of Congress Control Number: 2014953582

Mathematics Subject Classification: 91B30, 97M30

Springer Cham Heidelberg New York Dordrecht London

Printed on acid-free paper

Springer is part of Springer Science+Business Media (www.springer.com)

Preface

The book aims at providing the "basics" of health insurance techniques. The first chapters (Chaps. 1–3) explain the need for health insurance, the risks taken by an insurance company writing health insurance policies, and describe insurance products in this area (sickness insurance, accident insurance, critical illness covers, income protection, long-term care insurance, health-related benefits as riders to life insurance policies). Both individual and group policies are considered. Of course, issues of specific current interest, e.g., the design of recent insurance products, are also addressed.

A shift to general actuarial features of health insurance products follows (Chap. 4). Then, basic actuarial models are presented for sickness insurance (Chap. 5) and income protection, i.e., disability annuities (Chap. 6); a short introduction to actuarial models for long-term care insurance products is also provided.

As regards actuarial models, "basic" means that only the traditional equivalence principle is considered for premium and reserve calculations, whereas risk and solvency issues are briefly addressed. Actually, the book aims at offering a comprehensive introduction to the basic aspects of pricing and reserving, thus providing the reader with the technical tools needed to move ahead in the field of health insurance management.

The book has been planned and structured assuming as its target readers: advanced undergraduate and graduate students in Actuarial Sciences; graduate students in Economics, Business and Finance; professionals and technicians operating in insurance and pension areas.

It is assumed that the reader has attended courses providing basic notions of Financial Mathematics (interest rates, compound interest, present values, accumulations, annuities, etc.) and Probability (probability distributions, conditional probabilities, expected value, variance, etc). The mathematics has been kept at a rather low level: indeed, almost all topics are presented in a "time-discrete" framework, thus analytical tools like derivatives, integrals, etc., are not required. Some sections in which differential calculus has been used can be skipped without significant loss in understanding the basic calculation principles.

Although the book is "teaching" rather than "research oriented", many scientific and professional contributions to the health insurance technique have been included in the References, such as papers in scientific journals and conference proceedings, working papers, technical reports, etc. This material can provide substantial help, especially if, for specific topics, textbooks are not available, or not updated. Citations, together with some comments, are listed in a special section at the end of each chapter.

The logical structure and the contents of the book have successfully been tested in various teaching initiatives; in particular:

- courses, short courses, and seminars in several universities (University of Trieste, University of Louvain in Louvain-La-Neuve, European University at St. Petersburg, MIB School of Management in Trieste);
- seminars for professional bodies (Portuguese Institute of Actuaries, U.S. Society of Actuaries, Australian Institute of Actuaries, Polish Society of Actuaries), in the framework of CPD (Continuing Professional Development);
- CPD seminars in insurance companies (in Brussels and Lisbon).

If this book helps the reader to understand the basic technical features of the products in the manifold areas of health insurance, and stimulates the reader's interest in deepening his/her knowledge of more complex topics, such as the risk profile of health insurance portfolios, then it will have achieved its objective.

Acknowledgments

We would like to thank AG Insurance for their financial support. Part of this book has been developed within the framework of the AG Insurance Chair on Health Insurance at KU Leuven (chairholder: Jan Dhaene; co-chairholders: Katrien Antonio, Michel Denuit).

Trieste, July 2014 Ermanno Pitacco

Contents

Chapter 1
The Individual Perspective: The Need for Health Insurance

1.1 Introduction

This chapter aims to motivate the need for health insurance, as a tool to transfer the financial consequences of temporary or permanent alterations of individual health conditions, which can cause either health care-related costs or a negative impact on income because of reduced working capacity, or both.

Of course, the range of possible alterations in the individual health status is extremely broad, and consequently the range of possible financial impacts is broad as well, ranging from relatively small and routine expenses (that is, with high probability of occurrence) to huge costs, for example because of surgery needs.

Each individual can easily forecast routine expenses, and then can face these expenses via his/her ordinary income (or, possibly, via savings). Conversely, appropriate insurance covers are needed for low-probability high-cost events. Insurance products must be accordingly shaped, by stating policy conditions such as deductibles, franchises, exclusions, and so on.

The need for (private) health insurance differs considerably across various countries, being related to the amount of health care provided by the public sector. Public health care systems have developed over time, depending on local political, cultural and socio-economic traditions. As a result, different health care arrangements can be found in different countries.

At an individual level, the demand for health care reflects personal education and attitude towards health. As a consequence, the utilization of health services can significantly differ among individuals belonging to the same population.

At a population level, the age structure of the national population constitutes a critical issue in determining the total health care demand and, in particular, the need for public health care.

Since topics related to health care economics are not within the scope of this book, in what follows we simplify our terms of reference. In particular:

- we disregard population-related issues, such as the age structure of a national population;

E. Pitacco, *Health Insurance*, EAA Series,
DOI 10.1007/978-3-319-12235-9_1

- we refer to a generic individual whose age pattern of health-related needs can be considered, at least to some extent, as representative of a reasonable "average" in the population of interest;
- as regards funding arrangements to meet the costs of health insurance covers, we refer to individual health needs throughout time, disregarding possible sharing between public health systems and private health insurance.

1.2 Individual Cash-Flows

We focus on the following individual cash flows.

1. Inflows:

   a. earned income (wage, salary), during the working period, that is, over the age interval (x, ξ);
   b. pension annuity (and, possibly, purchased life annuities), from retirement age ξ onward, that is, post-retirement income.

2. Outflows consisting of health-related expenses:

   a. medical expenses (medicines, hospitalization, surgery, etc.);
   b. expenses related to long-term care (because of senescent disability).

We assume that the time profile of the individual inflows (that is, inflow 1.a followed by 1.b) can be represented as plotted in Fig. 1.1. Note that, for simplicity, we

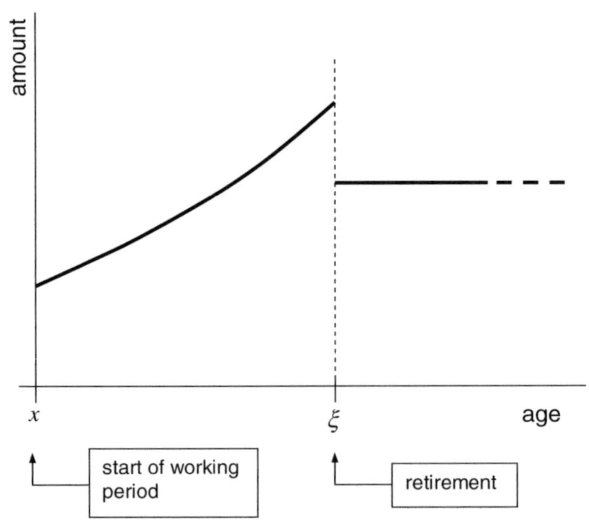

Fig. 1.1 The income profile: an example

have excluded possible interruptions of the regular income because of disability spells which can prevent the individual from working and getting the usual income. This income can partially be replaced by an appropriate "income protection" insurance, as we will see in Chap. 3. Further, we ignore inflation effects, so that the salary increase is due to career progression, whereas a flat profile is assumed for the post-retirement income.

1.2.1 The Time Profile of Health Care Costs

Remark In the following Figures, time profiles of health-related outflows are represented (together with possible funding arrangements). We stress that Figs. 1.2, 1.3, 1.4, 1.5, 1.6, 1.7 and 1.8 are about concepts, and do not necessarily correspond to real age patterns of health costs as these can result from statistical evidence. ■

A likely time profile of total expected costs for health care is shown in Fig. 1.2. (Of course, the scale adopted on the vertical axis is not the one used in Fig. 1.1.)

Health-related costs are not certain. More precisely, events implying health care expenses can be classified according to the probability of occurrence in a given time unit (say, a year) and the related financial impact. A rough (but rather meaningful) classification can be as follows:

1. high-probability low-cost events, which imply routine expenses (represented in Fig. 1.2 by the "distance" between the two curves);
2. medium-probability medium-cost events;
3. low-probability high-cost events.

This classification can help in understanding the variability of health-related costs, with significant contributions to such variability coming from items 2 and 3. At the

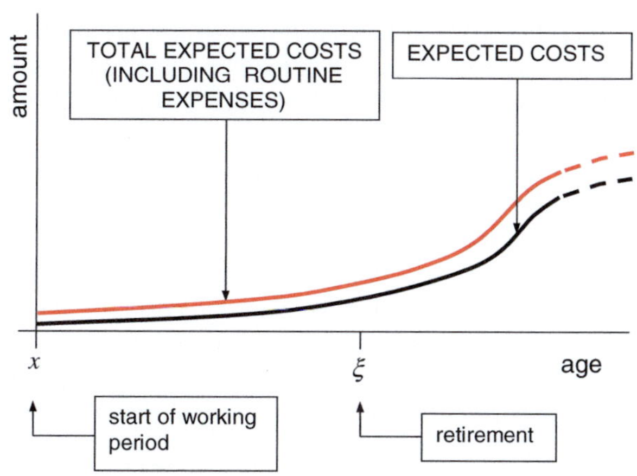

Fig. 1.2 Health-related expected costs

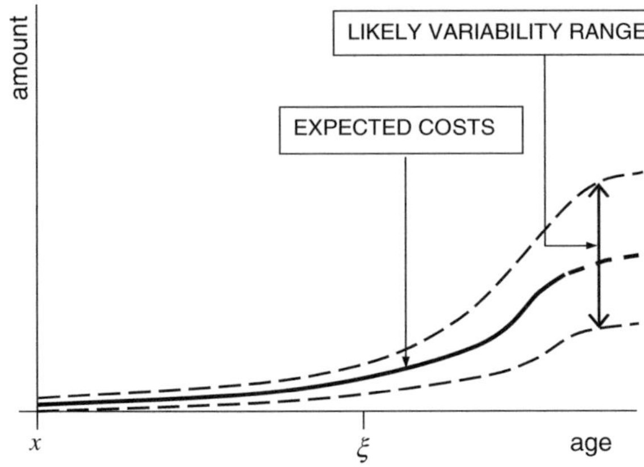

Fig. 1.3 Health-related costs: expected value and variability

same time, the above classification constitutes a tool for checking the appropriateness of different "strategies" for financing health-related expenses (as we will see in Sect. 1.3). In what follows we disregard routine expenses, by assuming for these expenses a variability close to zero.

The variability of (other) health-related costs can be expressed in terms of variance or standard deviation. Further, it can be represented in terms of confidence intervals at a given confidence level, so that a likely variability range can be plotted, as shown in Fig. 1.3.

1.2.2 Risk Transfer via Insurance

Because of their randomness, health care costs constitute a risk for each individual. We now assume that this risk can be transferred to another party, in particular to an insurance company. Individual costs will then be shared among the persons who constitute the insured pool (the insurer's "portfolio"), according to a cross-subsidy mechanism, i.e. the *mutuality* mechanism.

In order to transfer the risk, each individual will be charged a sequence of insurance *premiums*, whose progression throughout time can be arranged in various ways, depending in particular on the duration of the insurance cover.

The most "natural" arrangement consists in a sequence of periodic (say, annual) premiums, each premium being equal to the expected cost related to that period plus an appropriate *safety loading* charged by the insurer to face the risk arising from the randomness of the periodic health-related costs. The resulting periodic amount is frequently called the *natural premium* (and, as an alternative, the *risk premium*). In what follows, we assume the year as the time unit.

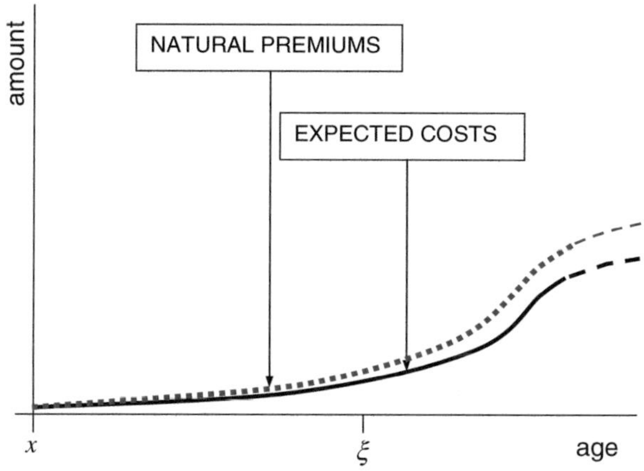

Fig. 1.4 Health-related expected costs and natural premiums (including safety loading)

The above arrangement is shown in Fig. 1.4. If the risk transfer is realized via a sequence of one-year insurance covers, the premium arrangement we have now described is the only feasible one.

The natural premiums must be paid as long as the individual is alive, and no accumulation process is implied (whatever the duration of the insurance cover may be), since each periodic premium meets the cost related to that period. Further, natural premiums increase as the individual age increases. The following aspects should in particular be stressed.

- From the insurer's point of view, no risk is originated by the randomness of the individual lifetime.
- From the individual's perspective, natural premiums at old and very old ages are very high, compared to the income profile (whereas they can be appropriate at young ages, in relation to the likely income).
- There is no guarantee that the insurer will provide the coverage over the whole lifespan, either because of the individual's very old age, or because a very high amount of health-related costs has been experienced by the insured and reimbursed by the insurer.

High amounts of premiums at very old ages can be avoided by resorting to premium arrangements based on some "leveling" principle. It is understood that any premium leveling can only be implemented in the presence of a multi-year duration of the insurance cover.

1.2.3 Temporary Insurance Covers

A health insurance cover can be realized via a sequence of m-year temporary covers (for example, with $m = 10$). Such an arrangement is rather common in several insurance markets. The sequence of temporary covers can embrace the whole lifespan.

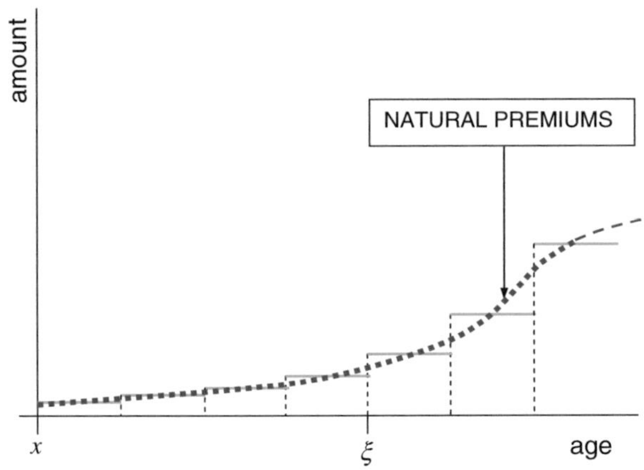

Fig. 1.5 Temporary covers: natural premiums and level premiums

Each temporary cover can be financed, in particular, either via natural premiums or via level premiums payable for the whole duration of the cover itself. An example of the latter arrangement is shown in Fig. 1.5. The related actuarial problems will be dealt with in Sect. 5.3. Here we only focus on the following features.

1. Assumptions about the insured's lifetime (which, of course, is random) are needed in relation to each m-year temporary cover, in order to calculate the amount of the related level premiums (as we will see in Sect. 5.3.2).
2. Level premiums are, in each temporary cover, initially higher and then lower than the relevant natural premiums; more precisely, whatever the cover duration and provided that level premiums are payable throughout the whole duration, the amount of the level premiums is a weighted arithmetic mean of the natural premiums. It follows that:

    a. a *reserving process* must be implemented by the insurer, in relation to each temporary cover, so that amounts exceeding the costs represented by natural premiums can be used while level premiums are not sufficient to cover the costs (see Sect. 5.3.3);
    b. in each time interval, premiums initially higher and then lower than the natural premiums are charged to the insured; however, level premiums are close to natural premiums in those age intervals in which the increase in natural premiums is small;
    c. the amounts of level premiums are anyway high at old and very old ages.

1.2.4 A Lifelong Insurance Cover

Assume that the cover is provided by an insurance product with a lifelong duration. Figure 1.6 shows the related premium arrangement based on *lifelong level premiums*; the amount of the annual level premiums is represented by the solid line.

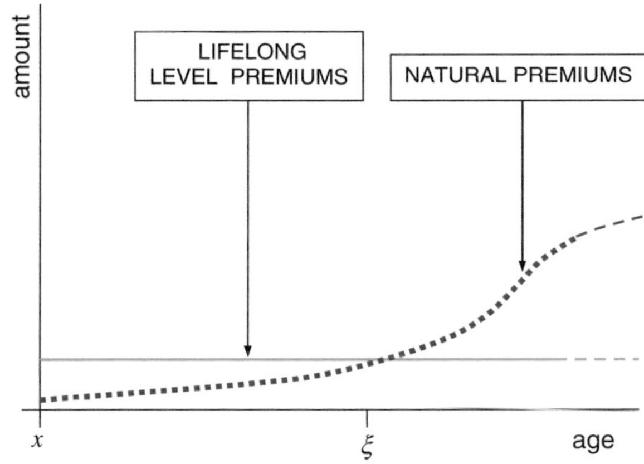

Fig. 1.6 Lifelong cover: natural premiums and lifelong level premiums

It is worth singling out the following aspects.

1. The calculation of level premiums must rely on biometric assumptions about the insured's random lifetime (as in the case of temporary covers with level premiums, see Sect. 1.2.3); hence, risks related to the insured's lifetime are taken by the insurer, as we will see in Sect. 2.3 (and the related impact may be much heavier than in the case of temporary covers).
2. Leveling of course results in premiums higher than the natural premiums in a first period, and lower later on. Then:

    a. an accumulation, or *reserving process*, must be implemented by the insurer;
    b. from the insured's point of view, leveling over the whole lifespan results in amounts more acceptable at very old ages (although acceptability depends on the post-retirement income); conversely, the insured is initially charged premiums which are much higher than the natural premiums.

The calculation of level premiums can also be implemented assuming that premiums are payable over a limited time interval. Then, *temporary level premiums* are determined. A reasonable term for premium payment is given by the retirement date, so that premiums are due up to the retirement age ξ. See Fig. 1.7.

We note the following aspects (see the corresponding comments concerning lifelong premiums).

1. Assumptions about the insured's lifetime are also needed in the case of temporary level premiums.
2. The temporary level premiums are much higher than the natural premiums related to the premium payment period; it follows that:

    a. a reserving process must be implemented by the insurer, resulting in accumulated amounts higher than those in the lifelong premium arrangement;

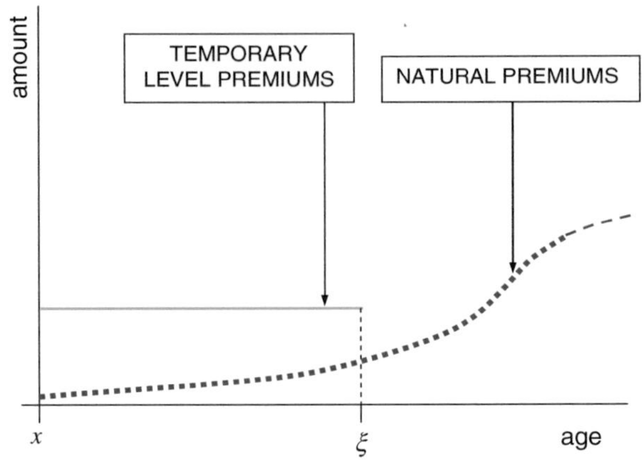

Fig. 1.7 Lifelong cover: natural premiums and temporary level premiums

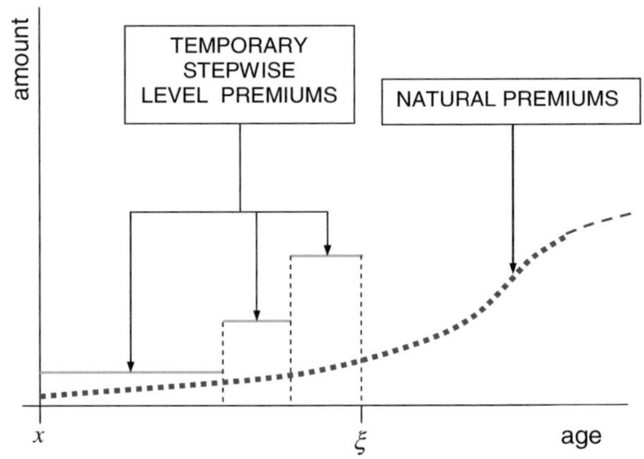

Fig. 1.8 Lifelong cover: natural premiums and temporary stepwise level premiums

b. the insured is initially charged very high amounts (compared to the natural premiums), which might be not appropriate given the initial income.

The disadvantage under point 2.b can be removed by resorting to *temporary stepwise level premiums*; see Fig. 1.8. This premium arrangement allows a high degree of flexibility (even if premiums are limited to retirement age ξ). So, the specific requirement of consistency between the premium levels and the income profile can be met.

Assumptions about the individual lifetime are also required in the case of temporary stepwise level premiums, and a reserving process must be implemented by the insurer.

Actuarial problems related to premium calculation for lifelong insurance covers will be dealt with in Sect. 5.5.

Remark So far, our comments have been restricted to aspects related to premium arrangements, for both temporary and lifelong covers. However, other characteristics of one-year, multi-year and lifelong insurance products must be analyzed in depth. These issues will specifically be addressed in Chap. 3. ■

1.3 Financing Health-Related Expenses

Various cover and premium arrangements have been introduced and commented on in Sects. 1.2.2 and 1.2.3, assuming an insurance-based financing of health-related costs. However, health-related costs (to the extent they are not covered by the public health care system) can be financed in several ways.

Table 1.1 summarizes the main features of three basic financing "strategies", with various implementations for the insurance strategy:

1. *out-of-pocket*, which means relying on ordinary inflows (e.g. salary) and available household assets;
2. *savings*, i.e. building up a fund via a progressive accumulation of resources, and withdrawing from the fund to meet expenses;
3. *insurance*, that is, arranging a (partial) transfer of the financial consequences of health-related risks by purchasing an insurance cover, and in particular:
   - 3.1 *one-year* covers, in which case the risk transfer throughout time requires a sequence of insurance covers;
   - 3.2 a sequence of *multi-year* covers, and possibly a *lifelong* cover, so that one insurance cover realizes the risk transfer over several years; as regards premiums, the following arrangements can be considered:
     - 3.2.1 *natural premiums*, each premium being related (in particular proportional) to the relevant expected annual cost, and hence increasing as the age increases;
     - 3.2.2 *level premiums* (and, possibly, *stepwise level premiums*), so that an "averaging" over time is realized.

Table 1.1 Financing health-related costs

			Pre-funding	Pooling
1 Out-of-pocket			No	No
2 Savings			Yes	No
3 Insurance	3.1 One-year		No	Yes
	3.2 Multi-year	3.2.1 Natural premiums	No	Yes
		3.2.2 Level premiums	Yes	Yes

Table 1.2 Strategies for financing health-related costs

Health-related event		Appropriate financing strategy
Probability	Cost	
High	Low	Out-of-pocket
Medium	Medium	Savings
Low	High	Insurance

The main features which characterize the above strategies are the following ones.

- *Pre-funding* means spreading health-related costs over time.
- Thanks to *pooling*, the total cost, in a given period, related to the members of the pool (an insurance portfolio in particular) is shared among the members themselves.

The interpretation of Table 1.1 is then straightforward. Some comments follow.

If the out-of-pocket strategy is implemented (or, more realistically, to the extent this strategy is implemented), the individual cannot rely on any re-distribution of health-related costs, that is, neither spreading over time nor sharing among individuals. Hence, this strategy can be appropriate for health-related events which only imply routine expenses, e.g. high-probability low-cost events in particular.

A savings-based strategy allows the costs to be spread over time, while the individual continues to be charged for his/her own costs. This strategy can comply with the characteristics of medium-probability medium-cost events.

Insurance strategies, whatever the duration of the insurance policy and, in the case of multi-year covers, whatever the premium arrangement, imply risk pooling and hence sharing of health-related costs among members of the pool (the insurance portfolio) according to the *mutuality mechanism* (which is the "raison d'être" of any insurance arrangement). This strategy is appropriate in particular for low-probability events which imply high costs.

Table 1.2 shows the most appropriate financing strategies for some typical probability-cost combinations. Of course, other possible combinations should also be considered, allowing in particular for the specific health care needs.

1.4 Suggestions for Further Reading

Various aspects of health insurance are dealt with in several textbooks; the reader can refer in particular to Black and Skipper (2000) (which provides useful and detailed information on insurances of the person in general, including health insurance), Bartleson (1968) and O'Grady (1988). Interesting references are also given by Bernet and Getzen (2004) and Szuch (2004).

Age patterns of individual health-related costs are analyzed in Yamamoto (2013).

Chapter 2
The Insurer's Perspective: Managing Risks

2.1 The Risk Transfer Process

The transfer of risks from individuals to an insurer implements a "transformation" of the risks themselves. The main aim of this section is to provide the reader with an insight into such a transformation.

2.1.1 The Risks Transferred to the Insurer

A number of definitions have been proposed for the term "risk", some of which belong to common language, whereas others relate to more specific business language, and the language of insurance business in particular.

As far as health-related events are concerned, we can define risk as the financial consequence of any temporary or permanent alteration of the individual's health conditions, which can cause either health care costs or a negative impact on income because of reduced working capacity, or both. Some examples follow:

- expenses for medicines prescribed by a physician;
- expenses for hospital stays;
- expenses for rehabilitative care;
- loss of income because of temporary (or permanent) disability.

As is clear from the examples, the above definition is rather generic. Indeed, several aspects of the possible health-related event should be specified. For instance, the event may last for either a short or a long period of time. In the latter case, the duration of the event can imply recurrent expenses, or a long-lasting loss of income.

We will not go into details in this section. More specific definitions will be provided in later sections when we deal with the various health insurance products which can provide financial protection in the face of health-related events.

E. Pitacco, *Health Insurance*, EAA Series,
DOI 10.1007/978-3-319-12235-9_2

Whatever the specific event, it is evident that the related risk is a *pure risk*, as it necessarily has a negative impact either on the income or on the wealth of an individual.

Health-related risks can be transferred to an insurer, against payment of a premium, or a sequence of periodic premiums. If one of the events specified in the insurance contract occurs, then the insurer pays the benefit, whose amount is determined according to the policy conditions. Various types of (monetary) benefits will be defined in Sect. 3.2.3.

2.1.2 The Risk Transformation

The insurance company acts as an *intermediary* in the risk transfer process. More precisely, a threefold intermediation role is played by the insurer:

- administrative;
- technical;
- financial.

The *administrative* intermediation consists in collecting premiums, issuing the insurance contracts, receiving the applications for benefits, settling the claims, and paying the benefits.

Although the importance of the administrative intermediation should not be underestimated (in particular as regards the claim settlement step), the *technical* intermediation is much more important as regards the management of the pool of risks. The role of the insurer as the technical intermediary consists in managing the mutuality within the pool of risks, namely the *portfolio*, providing the guarantee of paying the stated benefits whatever the number and the amounts of the claims, and hence taking the related risk (for more details, see Sect. 2.1.3).

Further, a *financial* intermediation is carried out when multi-year covers are involved, financed either by level premiums or by a single premium. In these cases, the insurer has to manage the funds over time, i.e. the reserves, originated by collecting the premiums (see Sects. 1.2.3 and 1.2.4).

As regards the technical intermediation, it is worth noting that, by managing the pool of risks, the insurer "transforms" the set of individual pure risks into a *speculative* risk, that is, the net result of the portfolio. The risk transformation can easily be described by referring to one-year insurance covers, e.g. providing medical expense reimbursement.

Let $X_1, X_2, \ldots, X_n$ denote the random benefits paid by the insurer to the n insureds who constitute the portfolio, and let $X^{[P]}$ denote the random total payout, i.e.

$$X^{[P]} = X_1 + X_2 + \cdots + X_n. \tag{2.1.1}$$

The generic insured j ($j = 1, 2, \ldots, n$) is charged a premium Π_j, so that the insurer cashes the amount

$$\Pi^{[P]} = \Pi_1 + \Pi_2 + \cdots + \Pi_n. \tag{2.1.2}$$

Disregarding expenses and the effect of interest over the year, the random net result is defined as follows:

$$Z^{[P]} = \Pi^{[P]} - X^{[P]}. \tag{2.1.3}$$

Assume that each premium contains an appropriate safety loading; then

$$\Pi_j > \mathbb{E}[X_j]; \quad j = 1, 2, \dots, n, \tag{2.1.4}$$

where $\mathbb{E}[X_j]$ denotes the expected value of the payment to the generic insured j. We then find:

$$\Pi^{[P]} > \mathbb{E}[X^{[P]}] \tag{2.1.5}$$

and finally:

$$\mathbb{E}[Z^{[P]}] > 0. \tag{2.1.6}$$

Hence:

- the expected net result is positive (and is equal to the total safety loading included in the premiums) and constitutes the portfolio profit margin; we denote this profit margin by $m^{[P]}$; thus

$$m^{[P]} = \mathbb{E}[Z^{[P]}]. \tag{2.1.7}$$

- $Z^{[P]}$ can, of course, take either positive or negative values (which respectively represent either *profits* or *losses*) depending on the number and the amounts of claims in the portfolio (see Fig. 2.1).

The transformation of the individual pure risks X_j ($j = 1, 2, \dots, n$) into the speculative risk $Z^{[P]}$ is formally represented by Eqs. (2.1.1)–(2.1.3).

Despite the safety loading $m^{[P]}$, the probability of a loss, $\mathbb{P}[Z^{[P]} < 0]$, can be rather high (see Fig. 2.1). To lower this probability, several risk management tools are available to the insurer (see Sect. 2.2), besides a raise in the safety loading, provided

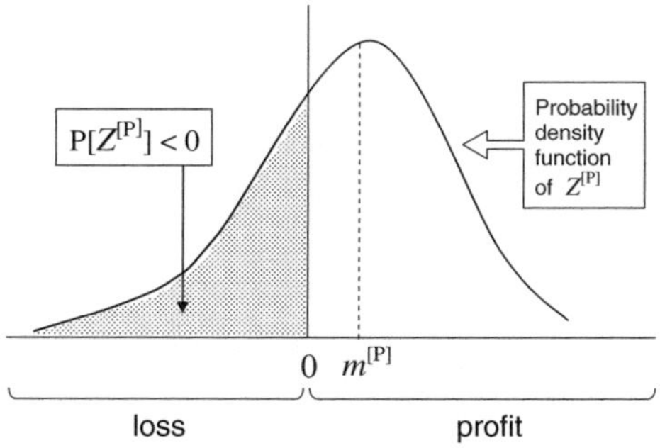

Fig. 2.1 Probability distribution of the net result from the portfolio

that this complies with obvious market constraints. Raising the safety loading (from $m^{[P]}$ to $m'^{[P]}$) determines a shift in the probability distribution of the net result $Z^{[P]}$ (see Fig. 2.2), and of course a decrease in the probability of a loss.

Moving from the individual pure risks X_j to the speculative risk $Z^{[P]}$ is one feature of the risk transformation process. The other important feature is the reduction of the *relative riskiness*. For a generic random variable Y, the relative riskiness can be quantified in terms of the *coefficient of variation*, $\mathbb{CV}[Y]$, also called the *risk index*, defined as follows:

$$\mathbb{CV}[Y] = \frac{\sigma[Y]}{\mathbb{E}[Y]}, \tag{2.1.8}$$

where $\sigma[Y]$ denotes the standard deviation of the random variable Y.

It can be proved that the relative riskiness of the portfolio net result $Z^{[P]}$, expressed by $\mathbb{CV}[Z^{[P]}]$, decreases as the portfolio size n increases, provided that some conditions concerning the random variables X_j are fulfilled. In particular, for a portfolio consisting of just one individual risk X_1, we trivially find $\mathbb{CV}[Z^{[P]}] = \mathbb{CV}[X_1]$, whereas (under appropriate conditions) we have:

$$\lim_{n\to\infty} \mathbb{CV}[Z^{[P]}] = 0. \tag{2.1.9}$$

In other words: the larger the portfolio size, the stronger is the reduction of the relative riskiness, namely the better is the *diversification*.

Remark Independence among individual risks plays an important role in achieving the diversification effect. However, the independence assumption should be questioned when catastrophic events, such as pandemics, are considered. For a more detailed discussion of risk diversification via pooling, in analytical and numerical terms, the reader should refer to the textbooks cited in Sect. 2.4. ■

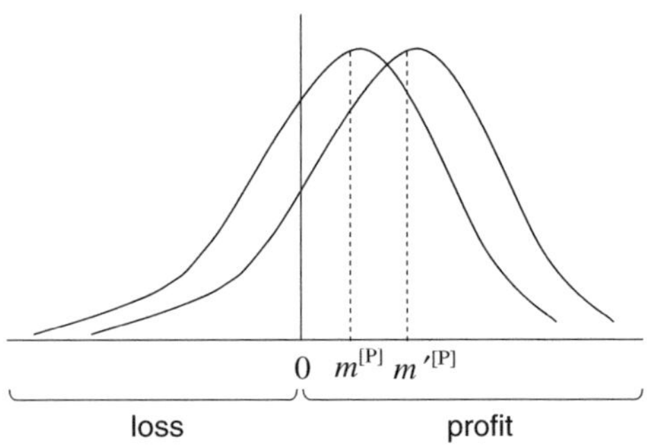

Fig. 2.2 Shift in the probability distribution of the net result

2.1.3 Insurer's Risk: Causes, Factors, Components

For a given type of health insurance cover, the possible outcomes of the random variables X_j (and the related probability distributions) are affected by several *risk causes*, also called *risk sources*. Of course, the outcomes of the X_j's contribute in determining the portfolio net result $Z^{[P]}$. Some examples follow:

- the morbidity among the insureds;
- the policyholders' behavior, and in particular the personal attitude of each policyholder towards health, which impacts on the "utilization" of the insurance cover;
- the mortality of the insureds, particularly relevant in the case of multi-year (and especially lifelong) covers.

Other risk causes directly affect the net result $Z^{[P]}$, for instance:

- the investment performance, in multi-year insurance covers financed via level premiums (or via a single premium) which imply a reserving process and hence the investment of the related assets (see the financial intermediation addressed in Sect. 2.1.2).

The various risk causes (and, in particular, the outcomes of the X_j's) impact on the net result $Z^{[P]}$. However, a more or less severe impact is a consequence of several *risk factors*, which can either raise or lower the effect of some risk causes on the portfolio result.

Some examples of risk factors follow (see also Fig. 2.3):

- a larger portfolio size improves the diversification effect, by reducing the relative riskiness of the portfolio result (as mentioned in Sect. 2.1.2);
- a small number of policies with (relatively) very high limit values, or sums assured, can jeopardize the diversification effect, at least to some extent; hence, the distributions of sums assured affects the riskiness of the portfolio net result;

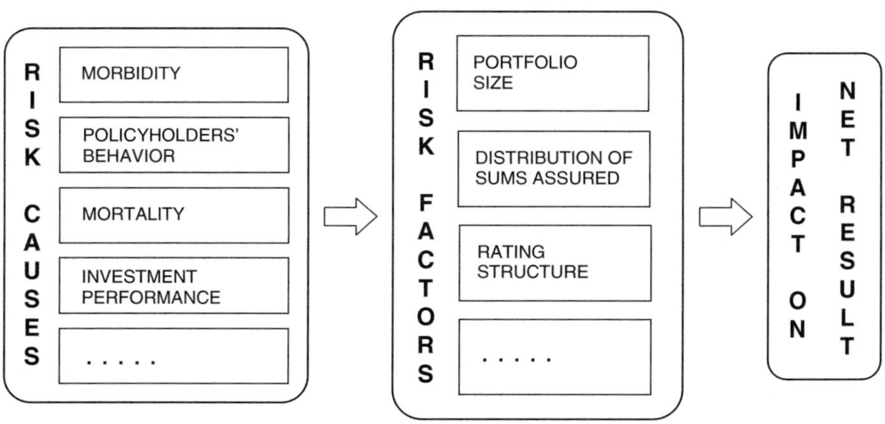

Fig. 2.3 From risk causes to portfolio net result

- of course, the rating structure adopted for premium calculations has a direct impact on the premium inflow $\Pi^{[P]}$ and hence on the net result $Z^{[P]}$.

As noted in Sect. 2.1.2, the net result $Z^{[P]}$ can take values far (and even very far) from its expected value $\mathbb{E}[Z^{[P]}] = m^{[P]}$, that is, the expected profit. In order to understand the reasons which can lead to a result significantly different from its expected value, an in-depth analysis is required.

For simplicity, we refer to fixed amount benefits provided by accident insurance in the case of total and permanent disability (see Sect. 3.3). We also assume that the sum assured, x, is the same for all the n policies in the portfolio. It follows that the total payout can be expressed as follows:

$$X^{[P]} = Kx, \tag{2.1.10}$$

where K denotes the random number of claims in the portfolio. Further, we suppose that the same probability of accident, p, is assigned to all the individuals. The expected value of the total payout is then given by:

$$\mathbb{E}[X^{[P]}] = \mathbb{E}[K]x = npx. \tag{2.1.11}$$

Let k denote the actual number of claims in the portfolio. If $k = np$, or, in terms of (relative) frequency, if $\frac{k}{n} = p$, the outcome of the total payout is given by npx, and hence coincides with its expected value. The portfolio result is then a profit, exactly equal to the expected profit $\mathbb{E}[Z^{[P]}] = m^{[P]}$ (see Eqs. (2.1.5)–(2.1.7)).

This (ideal) situation is represented by Fig. 2.4, in terms of the behavior of the (relative) actual frequency $\frac{k}{n}$ throughout time.

Conversely, we may find that $f \neq p$, at least in some years, and clearly our concern is for the case $f > p$. Figure 2.5 sketches three portfolio stories in which we find that, in various years, we have $f \neq p$. Reasons underlying this inequality may be quite different in each of the three stories.

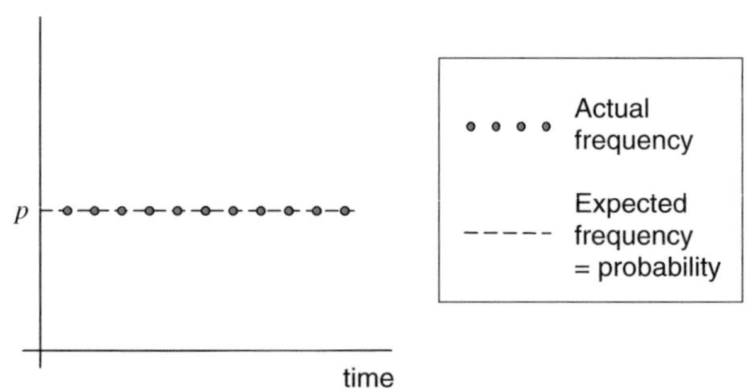

Fig. 2.4 Behavior of the relative frequency: the "ideal" situation

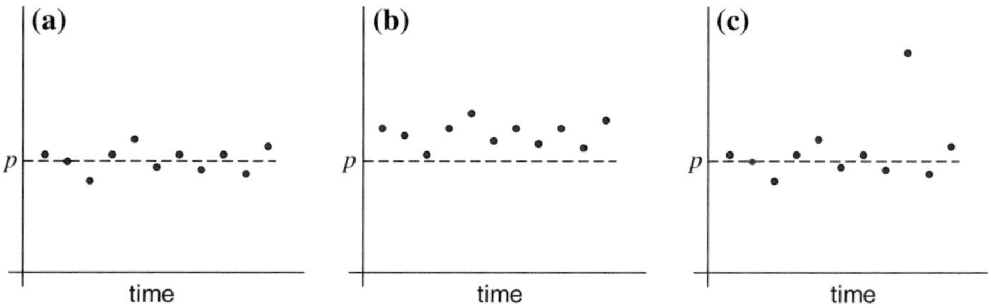

Fig. 2.5 Behavior of the relative frequency: more realistic situations

- In Fig. 2.5a, we see that the observed claim frequency randomly fluctuates around the probability, namely around the expected frequency p. This possibility is usually denoted as the *risk of random fluctuations*, or the *process risk*, or the *idiosyncratic risk*.
- In contrast, Fig. 2.5b depicts a situation in which, besides random fluctuations, we see "systematic" deviations from the expected frequency; this likely occurs because the assessment of the probability p does not capture the true nature of the insured risks. This possibility is usually called the *risk of systematic deviations*, or the *uncertainty risk*, referring to the uncertainty in the assessment of the expected frequency.
- In Fig. 2.5c, the effect of a "catastrophe", which causes a huge number of claims in a given year, clearly appears. This possibility is commonly known as the *catastrophe risk*.

Finally, Fig. 2.6 shows the combined effect of random fluctuations, systematic deviations and catastrophe risk.

Hence, three *risk components* have been singled out. All the components impact on the portfolio net result. However, the severity of the impact strongly depends on the portfolio structure, and the portfolio size in particular (as anticipated in Sect. 2.1.3).

Fig. 2.6 Behavior of the relative frequency: combined effect

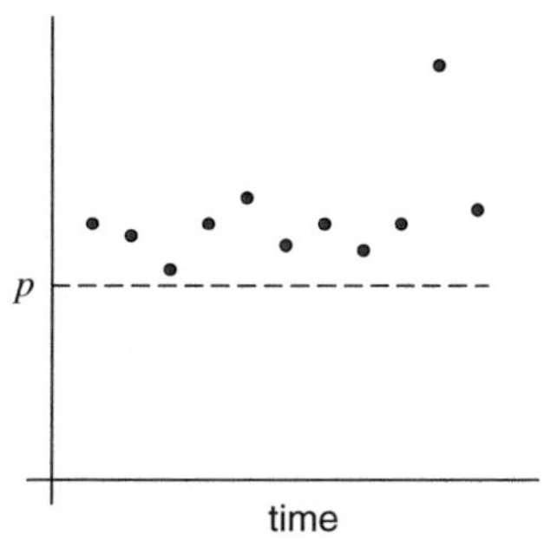

- The severity of the risk of random fluctuations decreases, in relative terms, as the portfolio size increases. This feature is a direct consequence of the risk pooling (see Sect. 2.1.2), and thus is commonly known as the *pooling effect*.
- The severity of the risk of systematic deviations is independent, in relative terms, of the portfolio size. Indeed, the systematic deviations affect the pool as an aggregate. Conversely, the total impact on portfolio results increases as the portfolio size increases.
- The severity of the catastrophe risk can be higher due, for example, to a high concentration of insured risks within a geographic area; as regards accident insurance in particular, a high concentration may refer to a group insurance policy (see Sect. 3.10) related to a large number of workers operating in the same workplace.

2.2 Risk Management Issues

The chapters dealing with (basic) actuarial models for health insurance (namely Chaps. 5 and 6) will focus on the traditional equivalence principle for premium and reserve calculations. However, models which are more complex than those based on the equivalence principle are needed for pricing insurance products that provide the policyholders with significant guarantees (hence charging the insurer with the related risks), and, more generally, for managing these products.

Actually, allowing in explicit terms for the impact of risks on portfolio results and the consequently needed "actions" implies the use of stochastic models and, in particular, of quantities like the Value at Risk (VaR) and the Tail Value at Risk (TVaR). While going into details regarding these issues is beyond the scope of this textbook, a glance at the main ideas underlying risk management principles can constitute a useful complement to one's knowledge of health insurance technical issues.

Remark In what follows we focus on some specific aspects of risk management. A wider perspective can be achieved by adopting the Enterprise Risk Management (ERM) approach, which provides guidelines for the management of risks, with an extensive range of applications (banking, insurance, commerce, industry, etc.). Actually, ERM is a holistic management process applicable in all types of firms and institutions. For a more in-depth analysis, with particular reference to the health insurance area, the reader can refer to the papers and reports listed in Sect. 2.4. ■

2.2.1 The RM Process

The implementation of the risk management (RM) principles takes place via the RM process.

The RM process essentially consists of the following steps (see Fig. 2.7).

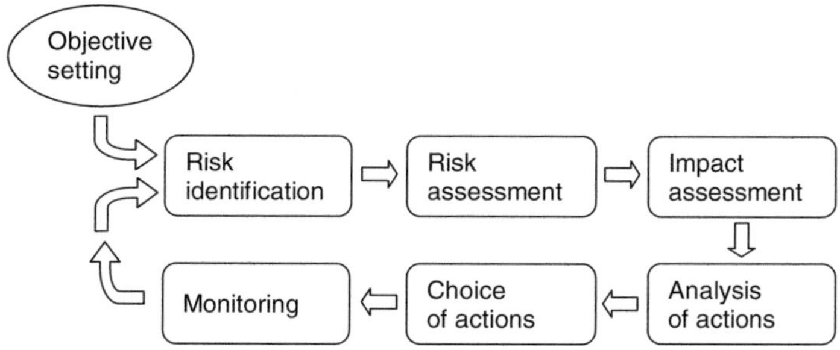

Fig. 2.7 Steps in the RM process

1. *Objective setting*. Any line of business (and hence health insurance portfolios in particular) aims at achieving given targets; some examples follow:
   - profit
   - value creation
   - solvency
   - market share
   - ...

   We will pay special attention to solvency and value creation in Sect. 2.2.2.
2. *Risk identification*. In this step risk "causes" (morbidity, disability, mortality/longevity, expenses, etc.) and risk "components" (random fluctuations, systematic deviations, catastrophic risk) are singled out (see also Sect. 2.1.3).
3. *Risk assessment*. Risk causes and risk components are expressed in quantitative terms via appropriate stochastic models (viz probability distributions).
4. *Impact assessment*. The impact of risk causes and risk components on results of interest (portfolio net result, cash-flows, profits, assets, etc.) is quantified in terms of probability distributions of results, and related typical values (expected values, variances, VaRs, etc.).
5. *Analysis of actions*. Costs and benefits related to possible insurer's actions (pricing of the products, reinsurance, capital allocation, etc.) are compared.
6. *Choice of actions*. Usually an appropriate mix of actions is chosen (e.g. combining reinsurance and capital allocation).
7. *Monitoring*. This step should involve both the results achieved by managing the insurance products and the statistical bases (e.g. morbidity, mortality/longevity) adopted when pricing the product.

It should be noted that the RM process is "never-ending". In fact, the monitoring step aims at checking the results of the adopted actions, and possibly suggesting a revision of the previously performed steps.

The RM process, as described above, refers to a given insurance product (or set of products), whose features (policy conditions and, in particular, guarantees and possible options) have already been defined in detail. When the insurer aims at launching a new product, the product design of course constitutes the first step

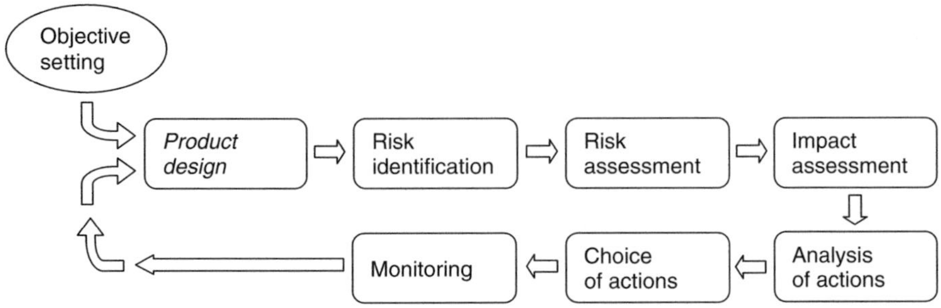

Fig. 2.8 Steps in the RM process, including product design

in the RM process (see Fig. 2.8). Then, the risk assessment and impact assessment steps might reveal a heavy risk exposure because of the product features, so that it is appropriate to include a possible re-design of the product among the actions.

Even for a given insurance product, the results provided by the monitoring phase can suggest a re-design of the product, e.g. in order to change some policy conditions and hence lower the related impact. Also in this case, it is appropriate to include the re-design of the product among the available actions.

Product design (and re-design as well) constitutes a critical step in the management of some health insurance products. An interesting example is provided by long-term care insurance (LTCI) covers (see Sect. 3.6) whose benefit structure suggests a product design involving an appropriate packaging of the specific LTCI benefits with lifetime-related benefits (some examples will be given in Sect. 3.6.3).

2.2.2 Capital Allocation: Solvency Versus Value Creation

RM principles are frequently understood as simply aiming at "risk mitigation" and, to the extent that risk impact cannot be further reduced, at finding appropriate tools to cover possible losses produced by a line of business, protecting the whole business against specific critical situations.

This interpretation is particularly inappropriate when applied to private insurance activity, given that its purpose is to gain profits by managing risks taken from other subjects, and this must be supported by a convenient amount of "risk appetite".

Indeed, a more appropriate interpretation of the RM principles and process should be driven by the objects of the activity (the insurance activity in particular) to which RM principles are applied, as clearly appears in the graphical representation of the RM process in Figs. 2.7 and 2.8.

All the objectives of the insurance activity should be carefully accounted for in the choice-of-actions step, which, as already noted, should consist in a convenient mix of "basic" actions.

Capital allocation is one of the basic actions, which aims at providing funds to cover possible portfolio losses. As is well known, capital allocation is mandatory to the extent that it is imposed by a specific solvency regime.

On the one hand, the higher the capital allocation, the higher the insurers' degree of solvency (which is, of course, one of the targets of the insurance risk management). On the other hand, allocating capital obviously implies a cost to the shareholders. The cost of shareholders' capital does not contribute in determining the insurer's profit. Conversely, *creation of value* relies on the possibility of covering all the cost, the cost of shareholders' capital included. Hence, the cost of capital allocation (besides the solvency requirements) should be carefully assessed when choosing the mix of RM actions.

2.3 Risks Inherent in the Random Lifetime

In the framework of risks related to the management of health costs, special attention should be placed on risks arising from the randomness of individual lifetimes. These risks should be carefully considered by both the individuals in the context of the household wealth management, and the insurers as regards the portfolio management when multi-year covers financed by level premiums are involved.

In what follows, we first focus on the so-called individual longevity risk (Sect. 2.3.1), then move on to the aggregate longevity risk (Sect. 2.3.2), which in particular arises from the uncertainty in future mortality trends (and thus constitutes an example of uncertainty risk, see Sect. 2.1.3). Finally, we address some issues concerning the management of lifetime-related risks (Sect. 2.3.3).

2.3.1 The Individual Lifetime

The lifetime of any given individual is, in mathematical terms, a random variable. Random fluctuations of the outcomes of individual lifetimes around the life expectancy (consistent with a given age-pattern of mortality) constitute the apparent consequence of this randomness.

The randomness in the individual lifetime is usually called the *individual longevity risk*. From a wealth management perspective, a consequence of the individual longevity risk is the risk of outliving the resources in the post-retirement period, in particular if an appropriate life annuity/pension is not available to the retiree.

As regards health-related costs, a lifelong insurance cover can face the impact of the individual longevity risk, provided that the insurance cover is financed in advance via a sequence of temporary periodic (possibly level) premiums (see Fig. 2.9). In this case, the insurer takes the risk inherent in the choice of the mortality assumption. This issue will again be addressed in Sect. 4.2.2.

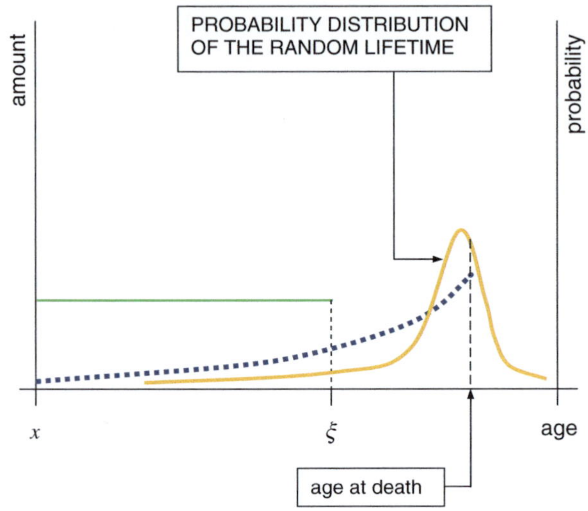

Fig. 2.9 Level premiums, natural premiums, and random lifetime

Various models have been proposed for the representation of the age-pattern of mortality, in terms of a life table or a survival function, which constitutes the mortality component of the *biometric assumptions*, also including, in the health insurance area, hypotheses about morbidity, disability, recovery, etc.

The age-pattern of mortality can be represented by various parametric models. In particular, the *Heligman–Pollard laws* take into account, thanks to their three terms, infant mortality, young-adult mortality, and senescent mortality. The first Heligman–Pollard law is given by the following expression:

$$\frac{q_x}{1-q_x} = \underbrace{A^{(x+B)^C}}_{\text{infant mortality}} + \underbrace{D e^{-E\,(\ln x - \ln F)^2}}_{\text{young-adult mortality}} + \underbrace{G H^x}_{\text{senescent mortality}}, \tag{2.3.1}$$

where x denotes the attained age and q_x the probability of dying between age x and $x+1$ for an individual aged x. The quantities $\frac{q_x}{1-q_x}$ are called the mortality odds.

Thanks to the numerous parameters, a high degree of flexibility is allowed by the law (2.3.1), so to represent a broad range of possible mortality assumptions. This law will be adopted in the numerical examples in Chaps. 5 and 6.

Example 2.3.1 Consider the Heligman–Pollard law with the following parameters:

$$\begin{array}{llll} A = 0.00054 & B = 0.017 & C = 0.101 & D = 0.00013 \\ E = 10.7200 & F = 18.67 & G = 1.464 \times 10^{-5} & H = 1.11000 \end{array}$$

We find in particular:

- life expectancy
  - at the birth: $\overset{\circ}{e}_0 = 79.412$
  - at age 40: $\overset{\circ}{e}_{40} = 40.653$
  - at age 65: $\overset{\circ}{e}_{65} = 18.352$
- Lexis point: $x^{[L]} = 85$

In Figs. 2.10 and 2.11 the probabilities q_x of dying between age x and $x+1$ for a person aged x, and their logarithms, are respectively plotted. Conversely, Fig. 2.12 shows the probabilities ${}_{x|1}q_0$ of dying between age x and $x+1$ for a newborn. □

2.3.2 The Longevity Dynamics

In many countries, mortality experience over the last decades shows some aspects affecting the shape of curves representing mortality as a function of the attained age. Figure 2.13 illustrates the moving mortality scenario for the Italian male population, in terms of curves of death, i.e. in terms of the number of people dying at age x, d_x, according to the usual demographic and actuarial notation; Fig. 2.14 provides the same information in terms of survival curves, i.e. in terms of the number of survivors at age x, ℓ_x, out of a notional cohort consisting of 100 000 newborns. The survival curves and the curves of deaths relate to various period mortality observations from 1881 to 2002 ("SIM t" refers to period observations on Italian males centered on calendar year t). Obviously, experienced trends also affect the behavior of other quantities expressing the mortality pattern, such as the life expectancy and the one-year probabilities of dying.

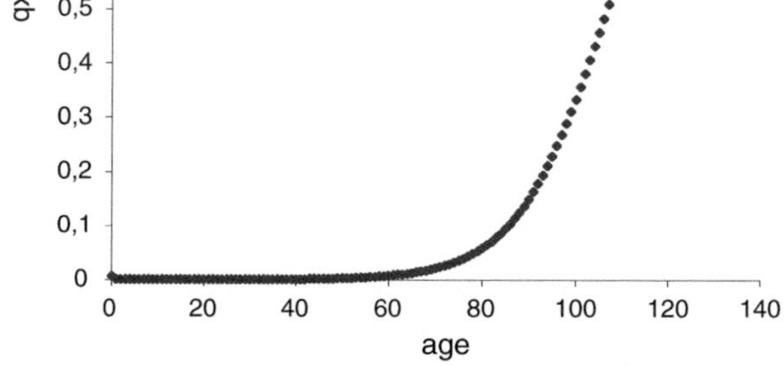

Fig. 2.10 The first Heligman–Pollard law: q_x

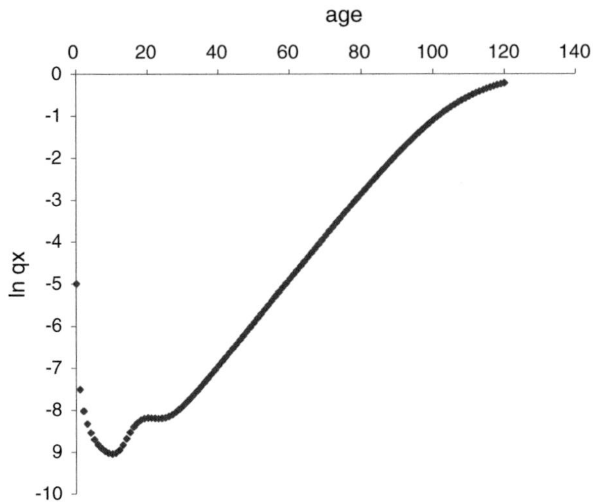

Fig. 2.11 The first Heligman–Pollard law: $\ln q_x$

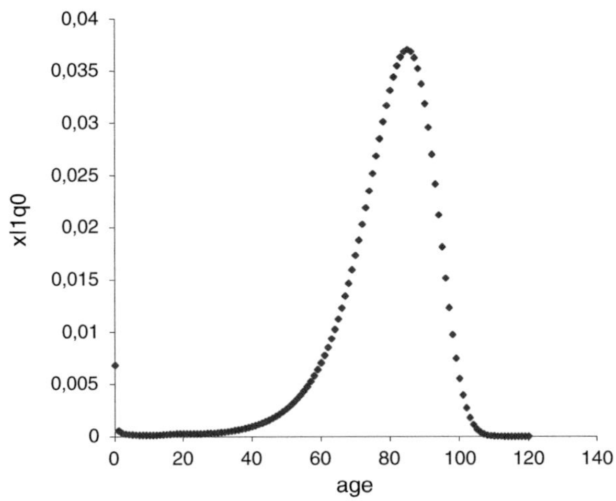

Fig. 2.12 The first Heligman–Pollard law: ${}_{x|1}q_0$

In particular the following aspects can be pointed out:

1. an increase in the life expectancy (at the birth as well as at old ages);
2. a decrease in the infant mortality, and in one-year probabilities of dying, q_x, in particular at adult and old ages.

Further, as regards the shape of the survival curves and the curve of deaths, the following aspects of mortality, common to many countries, can be singled out:

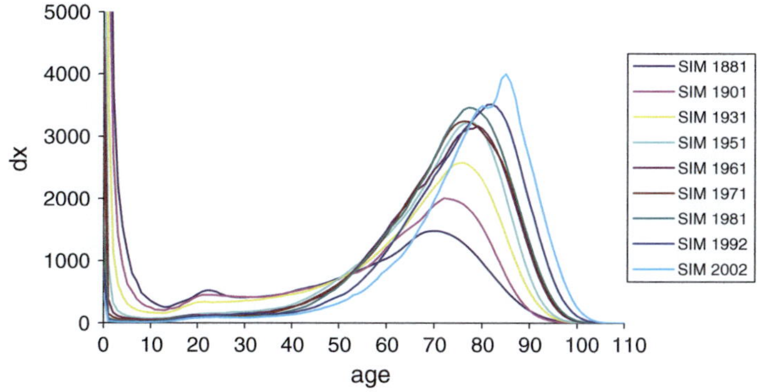

Fig. 2.13 Curves of death

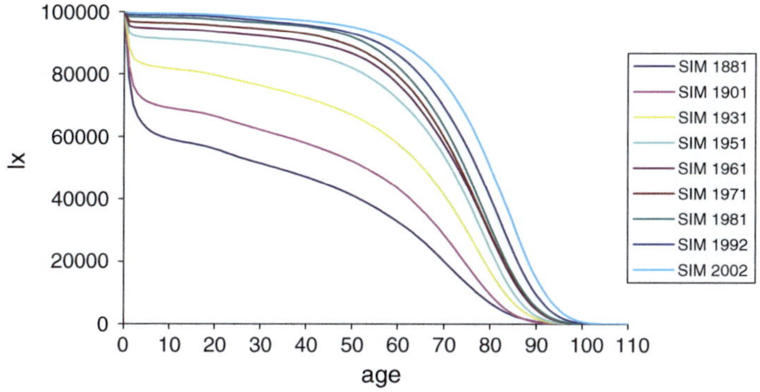

Fig. 2.14 Survival curves

3. an increasing concentration of deaths around the mode at old ages (that is, the Lexis point) of the curve of deaths is evident; so the survival function moves towards a rectangular shape, whence the term *rectangularization* to denote this aspect;
4. the mode of the curve of deaths (which, because of the rectangularization, tends to coincide with the maximum attainable age ω) moves towards very old ages; this aspect is called the *expansion* of the survival function;
5. higher levels and a larger dispersion of accidental deaths at young-adult ages (the so-called *young mortality hump*) have been observed more recently.

The progressive decline of human mortality, witnessed by a number of population statistics, suggests the rejection of the hypothesis of "static" mortality, which would lead to biased actuarial evaluations. Then, trends in mortality imply the use of "projected" life tables or mortality laws, for several purposes in life and health insurance, especially when long-term insurance products are concerned, for example life annuities and lifelong health insurance covers. A forecast of future mortality trend underpins the construction of projected life tables and mortality laws.

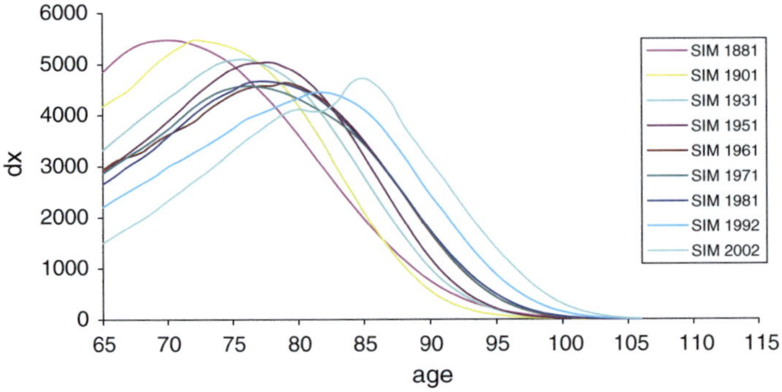

Fig. 2.15 Curves of death conditional on attaining age 65

It is also worth noting that an increasing dispersion of the individual lifetime conditional on attaining a given adult age (say, 65) has been observed in many countries, in contrast to the rectangularization over the whole lifespan. This aspect is shown in Fig. 2.15, in terms of numbers of people dying out of a notional cohort of people alive at age 65.

Whatever projected life table (or parametric law) is assumed for actuarial calculations, the future trend in mortality is random. The risk then arises that the average lifetime in a population (or portfolio) is larger than what has been expected according to the projected life table. This risk is called the *aggregate longevity risk.*

2.3.3 Managing the Longevity Risk

The individual longevity risk (see Sect. 2.3.1) and the aggregate longevity risk are the two components of what is named the *longevity risk* for short, that is, the risk that an individual lifetime or the average lifetime (in a given population) is higher than expected.

Splitting the longevity risk into the two components is important especially from the insurer's point of view. Indeed, the following features of the two components should be stressed.

- It is well known that the individual longevity risk (which is a "process risk", see Sect. 2.1.3) can be diversified via pooling. Referring to the total payout of an insurance portfolio, the relative impact of this component, which can be quantified by the risk index (or coefficient of variation), decreases as the size of the portfolio increases. If this size cannot be increased, (traditional) reinsurance transfers help in managing the individual longevity risk.
- The relative impact of the aggregate longevity risk is independent of the portfolio size (whereas the absolute impact on the portfolio payout increases as the portfolio size increases). Hence, no diversification is possible thanks to the pool

size or the traditional reinsurance arrangements. Alternative risk transfers must be implemented, as this risk component may have a dramatic impact on long-term products, viz life annuities and pensions, and on lifelong health insurance covers in particular.

An in-depth analysis of the longevity risk is beyond the scope of this book. The reader can refer to the relevant bibliographic suggestions in Sect. 2.4 and, as regards lifelong sickness insurance, in Sect. 5.6. Further, specific issues regarding the dynamics of disability-related life expectancy will be addressed in Sect. 6.16.3, and relevant bibliographic suggestions will be provided in Sect. 6.18.

2.4 Suggestions for Further Reading

General aspects of risk transfer to a pool and the specific role of the insurer in the pooling process are analyzed in Olivieri and Pitacco (2011).

The role of Enterprise Risk Management (ERM) in insurance and pensions is discussed in IAA (2009, 2011), whereas Orros and Smith (2010) and Rudolph (2009) specifically deal with ERM in the health insurance area; Orros and Howell (2008) in particular focusses on creation of value in health insurance.

Longevity issues (and the longevity risk in particular) are addressed in Pitacco et al. (2009); the interested reader should also refer to the numerous citations therein.

Chapter 3
Health Insurance Products

3.1 Introduction

In a broad sense, the expression *health insurance* denotes a large set of insurance products which provide benefits in the case of need arising from either accident or illness, and leading to loss of income (partial or total, permanent or non-permanent), and/or expenses (hospitalization, medical and surgery expenses, nursery, rehabilitation, etc.).

Health insurance, in its turn, belongs to the area of *insurances of the person*, where we find (see Fig. 3.1):

- life insurance (in a strict sense) and life annuities, which provide benefits depending on survival and death only, i.e. on the insured's lifetime;
- health insurance, which provides benefits depending on the health status and related financial consequences (and depending on the lifetime as well);
- other insurances of the person, whose benefits are due depending on events such as marriage, birth of a child, education and professional training of children, etc.

The shaded boxes in Fig. 3.1 represent insurance products commonly grouped under the label *protection*.

Health insurance products are usually shared between "life" and "non-life" branches according to national legislation and regulation.

Remark It is worth stressing that, in the insurance terminology, the meaning of several terms is not univocally accepted. As regards the particular field we are focussing on, a less broad meaning is frequently assigned to the expression "health insurance". For example, annuity-based products like those provided by disability insurance or income protection (see Sect. 3.5), or by long-term care insurance (see Sect. 3.6) are often excluded from the health insurance area, in which are conversely included products such as medical expense reimbursement or hospitalization benefits (see Sect. 3.4). ■

E. Pitacco, *Health Insurance*, EAA Series,
DOI 10.1007/978-3-319-12235-9_3

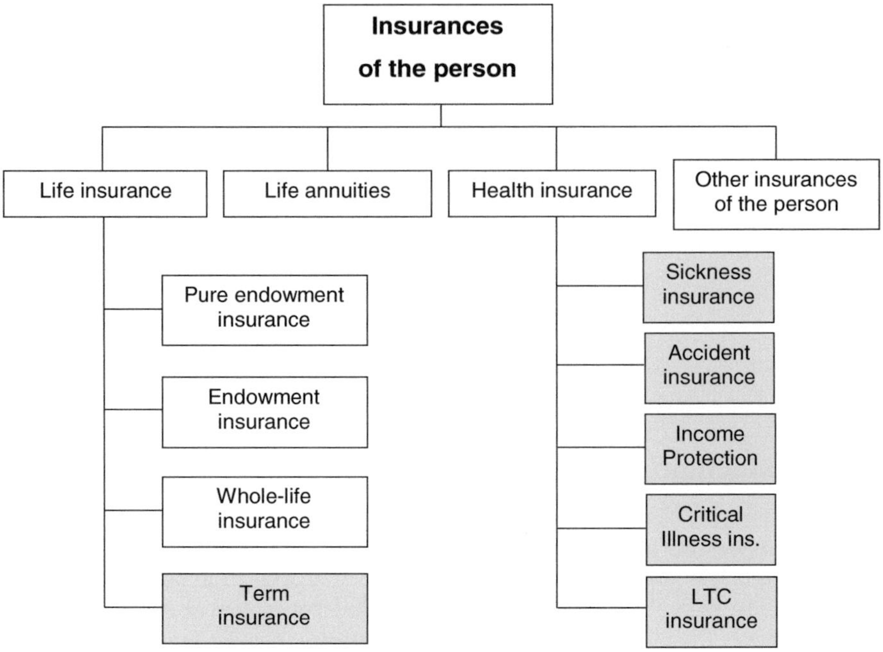

Fig. 3.1 Insurances of the person: basic products

3.2 Products and Types of Benefits

In this section a first insight into general aspects of (private) health insurance products is provided. An in-depth analysis will be presented in Sects. 3.3–3.7.

3.2.1 Main Health Insurance Products

Accident insurance covers a range of risks which may be caused by an accident (in particular, but not only, the risks of permanent disability and death). Various types of benefits can be included in the policy. A benefit frequently included consists of a lump sum paid in the case of permanent disability; the benefit amount is then determined as a function of the sum assured and the severity of the injury (i.e. the degree of disability). Accident insurance is described in Sect. 3.3.

Sickness insurance policies include medical expense reimbursement (that is, reimbursement, usually partial, of costs related to sickness or childbirth), and possibly hospitalization benefits (which consist of periodic payments during hospitalization periods, whose amounts are not related to actual expenses), as well as fixed-amount benefits in the event of temporary or permanent disability. See Sect. 3.4.

The term *disability insurance* denotes various types of covers, providing benefits in case of temporary or permanent disability. In particular, *income protection* (briefly *IP*) policies provide a periodic (usually weekly or monthly) income to an individual if he/she is prevented from working, and hence from getting his/her usual income, by sickness or injury. In the case of permanent disability, the benefit can consist of a lump sum instead of a sequence of periodic amounts. The main features of disability insurance products are described in Sect. 3.5, with special attention placed on IP policies.

Long-term care insurance (*LTCI*) provides the insured with financial support, while he/she needs nursing and/or medical care because of chronic (or long-lasting) conditions or ailments. Usually, annuity benefits are provided. See Sect. 3.6.

Critical illness insurance (*CII*), or *Dread disease insurance*, has a very limited extension of the coverage, which is defined via listing (rather than via exclusions). Diseases commonly covered are: heart attack, cancer, stroke, and coronary artery diseases requiring surgery. The benefit is a fixed lump sum. A CII cover frequently constitutes a rider benefit to a basic life policy including death benefit (typically a term insurance, or an endowment insurance) instead of being a stand-alone cover. CII products are described in Sect. 3.7. Other limited-scope products are described in Sect. 3.8.

3.2.2 The Policy Term

Health insurance may be provided by both *one-year* and *multi-year* covers, and, in particular, *lifelong* covers. For example, accident insurance is typically based on one-year policies, whereas income protection is usually provided by multi-year policies. Durations shorter than one year apply, for instance, to policies providing coverage of health risks related to travels and sojourns.

It is worth noting that a long-duration cover, and in particular a lifelong cover, can improve the "quality" of the insurance product from the policyholder's point of view. As regards sickness insurance, for example, an arrangement based on one-year covers (or even on longer temporary covers) does not provide the individual with any guarantee that the insurer will extend the coverage over the whole lifespan, either because the individual has attained a very old age, or because a very high amount of health-related costs has been experienced by the individual and reimbursed by the insurer (see also Sect. 1.2.2). Further, if specific underwriting requirements must be fulfilled at the time of renewal, a higher premium may be charged because of worsened health conditions.

Specific restrictions, which determine the actual coverage period, will be defined in Sect. 3.2.4.

One-year covers on the one hand and multi-year covers on the other require different actuarial structures, as we will see in Chap. 4, the latter also implying a life insurance-like approach and related biometric assumptions.

3.2.3 Monetary Benefits and Service Benefits

As regards the amounts of *monetary benefits* provided by health insurance products, we can recognize the following types.

Reimbursement benefits are designed to meet health costs (totally or partially), for example medical expenses. Hence, this category consists of *expense-related benefits*. Limitations such as deductibles and limit values are usually included in the policy conditions.

The amount of a *predefined benefit* is stated at policy issue, for example to provide an income when the insured is prevented from working by sickness or injury. In this category, we find both annuity benefits and lump sum benefits as well. See, for example, disability insurance products in Sect. 3.5. As regards the amount of the annuity or the lump sum, we have:

- *fixed-amount benefits*, which are independent of the severity of the health-related event and the possible consequent costs;
- *degree-related benefits* (or *graded benefits*), whose amount is linked to the severity of the health status expressed by some degree, e.g. the degree of disability; this arrangement is usually adopted in accident insurance policies (see Sect. 3.3) and in long-term care covers (see Sect. 3.6).

In the case of *service benefits*, a care service is provided by the insurer, relying on an agreement between care providers (e.g. hospitals) and the insurer. A special type of long-term care service benefit is provided in the US by the CCRCs (Continuing Care Retirement Communities; see also Sect. 3.6.5).

Figure 3.2 shows a classification of health insurance policies providing monetary benefits, which is based on the definition of the benefit amount (in particular: predefined benefit and expense-related benefit). We note that, when the benefit has an annuity-like structure (i.e. consisting of periodic payments), the related amount

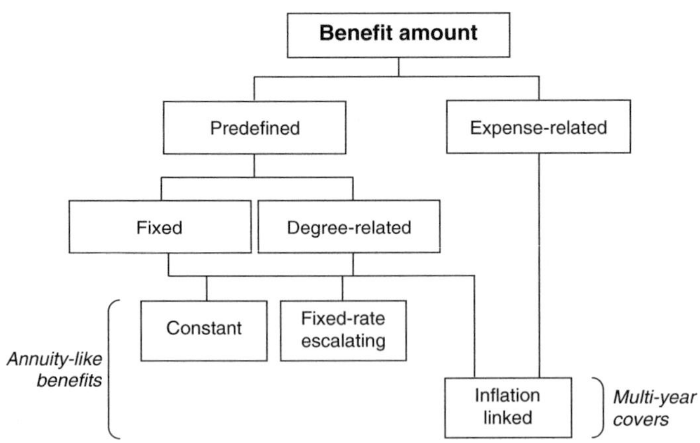

Fig. 3.2 Defining the amount of monetary benefits: a classification

can be either *constant* over time or *escalating*, e.g. in geometric progression with a fixed annual rate of increase. Further, an *inflation-linked* benefit can improve the quality of multi-year covers, both in the case of annuity-like benefits and in the case of expense-related benefits.

3.2.4 Policy Conditions

Several policy conditions are strictly related to the type of benefits provided by the health insurance products (for example: underwriting requirements, amount of deductible, allocation of the policy reserve in the case of early termination of the contract, etc.), and will then be analyzed while dealing with the various types of cover and the related actuarial models.

In this section we only address duration-related conditions, i.e. policy conditions which either define the coverage period (and relate to various health insurance covers) or the benefit payment period following the claim (and hence relate to annuity-like benefits, consisting in sequences of periodic payments). Some important duration-related conditions are represented in Fig. 3.3, where a generic disability claim in a multi-year cover is referred to.

The *insured period* (or *coverage period*) is the time interval during which the insurance cover operates, in the sense that a benefit is payable only if the claim time belongs to this interval. In principle, the insured period begins at policy issue, say at time 0, and ends at policy termination, say at time m. However some restrictions to the insured period may follow from specific policy conditions.

In particular, the *waiting period* is the period following the policy issue during which the insurance cover is not yet operating for sickness-related claims. Different waiting periods can be applied according to the category of sickness. The waiting

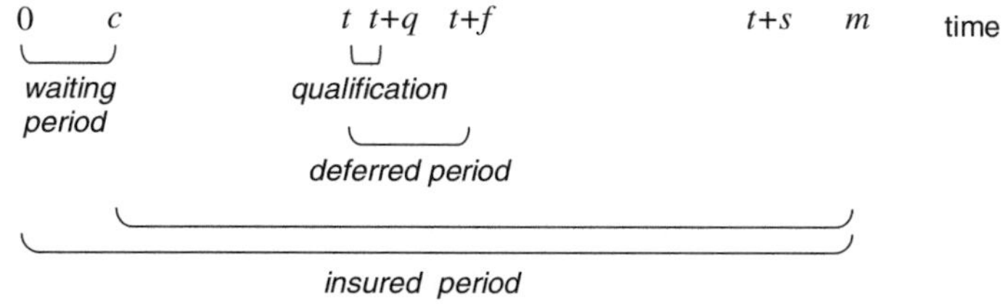

Fig. 3.3 Some policy conditions

period aims at limiting the effects of adverse selection. It is worth noting that, although the term waiting period is widely adopted, this time interval is sometimes called the "probationary" period (for instance in the US), while the term waiting period is used synonymously with "deferred" or "elimination" period (see below).

In many policies the benefit is not payable until the disability due to sickness or accident has lasted a certain minimum period called the *deferred period*. In sickness and accident covers the deferred period is usually rather short, mainly aiming at reducing the cost and hence the premium of the insurance products. Conversely, disability covers providing annuities are usually bought to supplement the (short-term) sickness benefits available from an employer or from the state, and hence the deferred period tends to reflect the length of the period after which these benefits reduce or cease. The deferred period is also called the (benefit-) *elimination period* (for example in the US). Note that if a deferred period f is included in a policy written to expire at time m, the insured period actually ends at policy duration $m - f$. Further, if the waiting period is c and the deferred period is f, the actual duration of the insured period is $m - c - f$, as regards sickness-related claims.

When a lump sum benefit is paid in case of permanent disability, a *qualification period* is commonly required by the insurer in order to ascertain the permanent character of the disability; the length of the qualification period would be chosen in such a way that recovery would be practically impossible after that period.

The *maximum benefit period* is the upper limit placed on the period for which benefits are payable (regardless of the actual duration of the sickness or the disability). In sickness insurance products providing daily benefits the maximum benefit period can be rather short, whereas it can be much longer in disability insurance covers, where the maximum number of years of annuity benefit payment may range from, say, one year to a lifetime. Different maximum benefit periods can be applied to accident disability and sickness disability. Note that if a long maximum benefit period operates, the benefit payment may last well beyond the insured period.

Another restriction to benefit payment may follow from the *stopping time* (from policy issue) of annuity payment. In disability annuities, the stopping time often coincides with the retirement age. Hence, denoting by x the age of the insured at policy issue and by ξ the retirement age, the stopping time r (from policy issue) is given by $r = \xi - x$.

When conditions such as the deferred period or a maximum benefit period are included in the policy, in case of recurrent disabilities within a short time interval it is necessary to decide whether the recurrences have to be considered as a single disability claim or not. The term *continuous period* is used to denote a sequence of disability spells, due to the same or related causes, within a stated period (for example six months). So, the claim administrator has to determine, according to all relevant conditions and facts, whether a disability is related to a previous claim and constitutes a recurrence or has to be considered as a new claim.

3.3 Accident Insurance

Accident is usually meant as an "unintended, unforeseen, and/or violent event, which directly causes bodily injuries". Accident insurance is also referred to as *personal accident insurance*.

3.3.1 Types of Benefits

Various benefits can be provided by accident insurance policies. We focus on the following.

Death benefit consists of a lump sum paid in the case the insured dies as a result of an accident. Clearly, this benefit is a fixed-amount benefit.

Permanent disability benefit consists of a lump sum paid in the case of dismemberment. Usually this is a degree-related benefit, because the amount paid is determined according to a benefit schedule. The benefit schedule first associates a disability degree to the severity of injury; then, the amount of the benefit is expressed as a percentage of the sum insured.

Figure 3.4 shows some examples of benefit grading, so that the benefit amount is a function of the severity of the injury. A *franchise deductible* is applied in case (b), a *deductible* with "adjustment" determines the benefit in case (c), whereas no deductible is applied in case (a).

Reimbursement of medical expenses can be provided by an accident insurance policy, in the case the expenses are related to a covered accident. Some related policy conditions are described in Sect. 3.4.2.

A *daily benefit during disability spells* can be paid as a fixed-amount benefit in the case of temporary disability caused by a covered accident. Usually the maximum benefit period ranges, according to policy conditions, from some months to one year.

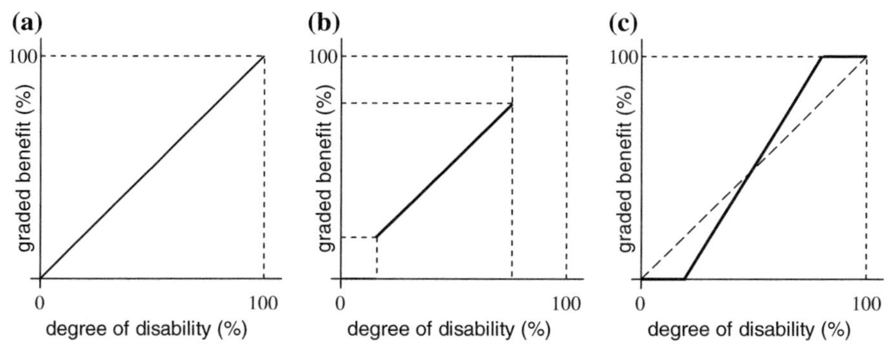

Fig. 3.4 Graded benefit as a function of the disability degree

3.3.2 Other Features

Accident insurance policies usually provide a one-year cover (but in the case the accident cover is a rider to a multi-year life insurance policy).

Among the policy conditions, it should be noted that a qualification period is usually applied in the case of permanent disability benefit.

Several exclusions can be stated in the policy conditions, which limit the range of covered accidents. Typically war-related accidents as well accidents related to illegal activities are not covered (unless otherwise specified in the policy). In some countries, the case of homicide as an accident is excluded from basic accident covers (while it can be explicitly added as a supplementary cover).

Special insurance plans are designed to cover specific types of needs. For example, *travel accident insurance* only covers accidents occurred while the insured is traveling. *Student accident insurance* covers accidents occurring while the student is engaged in school activities.

3.4 Sickness Insurance

Sickness insurance policies provide benefits in the event the insured becomes sick. The benefits provided by sickness insurance vary according to the type and the extension of the insurance policy. We focus on the most common benefits.

3.4.1 Types of Benefits

Reimbursement of medical expenses is the most important benefit provided by a sickness insurance policy. The benefit package can include various items:

- *hospital inpatient* cover includes all the services provided while the insured is hospitalized, including surgery, lab tests, drugs, etc.;
- *outpatient* cover relates to services provided in a physician's office and hospital outpatient setting, including minor surgery;
- further coverage includes the reimbursement of expenses related to lab tests, to physician-prescribed drugs, etc.

The *hospitalization benefit* consists of a fixed-amount daily benefit paid during hospital stays. This benefit is not expense-related.

A *daily benefit during sickness spells* can be provided as a fixed-amount benefit in the case of temporary disability caused by a covered sickness. Usually, the maximum benefit period ranges, according to policy conditions, from some months to one year.

Permanent disability benefit consists of a lump sum paid in the event of permanent disability caused by sickness.

3.4.2 Some Policy Conditions

In this section we only refer to sickness insurance policies providing reimbursement of medical expenses.

The *deductible* (also called *flat deductible*, or *fixed-amount deductible*) is a predefined amount that the insured has to pay out-of-pocket before the insurer will (partially) cover the remaining eligible expenses. Depending on the insurance product, the deductible can either refer to each single claim (sickness or injury), or to the policy period (e.g. the policy year). The deductible works as a disincentive for the insured to incur unnecessary medical expenses. Of course, the higher the deductible the lower the premium.

The *proportional deductible* (also called *fixed-percentage deductible*, or *coinsurance*) is the fraction of eligible medical expenses that the insured has to pay, after having met the flat deductible.

The *stop-loss* is the maximum amount the insured will pay out-of-pocket for medical expenses. It can be referred either to each single claim or to the policy period.

Once the above policy conditions have been stated, medical expenses are consequently shared between insured and insurer. The sharing can formally be described as follows. Assume that the policy conditions refer to each claim. We adopt the following notation:

x = generic expense amount;
D = flat deductible;
α = proportional deductible;
SL = stop-loss;
M = amount which depends on D, α, SL (see Eq. (3.4.3));
u = out-of-pocket payment;
y = reimbursement benefit paid by the insurer.

Of course $u + y = x$. We have, for $0 < \alpha \leq 1$:

$$u = \begin{cases} x & \text{if } x < D, \\ \alpha\,(x - D) + D & \text{if } D \leq x < M, \\ SL & \text{if } x \geq M, \end{cases} \tag{3.4.1}$$

$$y = \begin{cases} 0 & \text{if } x < D, \\ (1 - \alpha)\,(x - D) & \text{if } D \leq x < M, \\ x - SL & \text{if } x \geq M, \end{cases} \tag{3.4.2}$$

where

$$M = \frac{1}{\alpha}\big(SL - (1 - \alpha)\,D\big). \tag{3.4.3}$$

Of course, $M = SL$ if $\alpha = 1$, and $M = \dfrac{1}{\alpha}\,SL$ if $D = 0$ and $0 < \alpha \leq 1$.

The solid lines in Figs. 3.5 and 3.6 show the expense sharing, as defined by Eqs. (3.4.1) and (3.4.2) respectively.

Example 3.4.1 Assume: $D = 100, \alpha = 0.25, SL = 500$. We find (see Eq. (3.4.3)): $M = 1\,700$. Table 3.1 shows some examples of sharing according to Eqs. (3.4.1) and (3.4.2). □

Fig. 3.5 The out-of-pocket payment

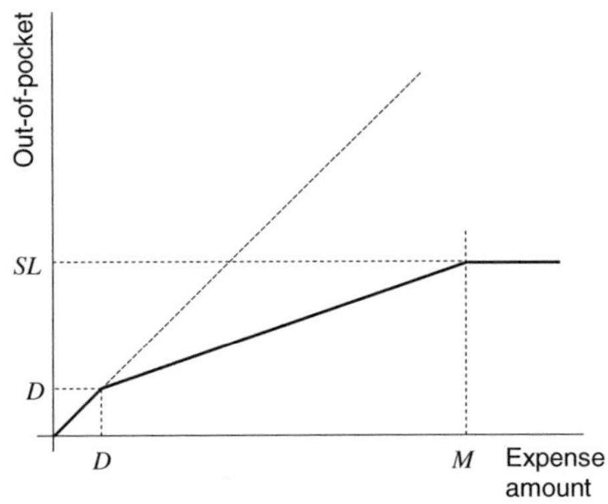

Fig. 3.6 The reimbursement benefit

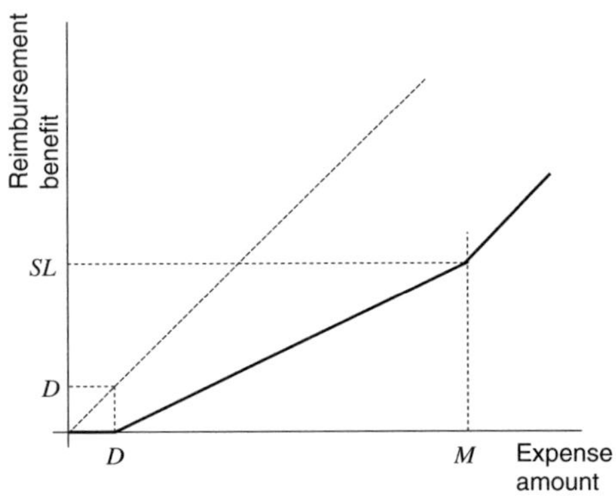

Table 3.1 Sharing medical expenses

Expense amount x	Out-of-pocket u	Reimbursement benefit y
50	50	0
300	150	150
900	300	600
1 800	500	1 300

3.4.3 Other Features

Underwriting requirements are usually applied in order to assess the individual health status at policy issue. The underwriting process may result in higher premium rates for *substandard risks*, that is, when poor health conditions are ascertained.

Guaranteed issue products are sickness insurance products which can be sold with little or no underwriting requirements. In this case, a higher premium is usually charged against the risk of adverse selection.

Waiting periods are commonly applied to limit possible adverse selection. A qualification period is also applied when a lump sum benefit in the event of permanent disability is paid.

A sickness insurance product can provide coverage to more individuals, in particular all the members of a family.

3.5 Disability Insurance

As noted in Sect. 3.2.1, the term *disability insurance* denotes various types of covers, providing benefits in case of temporary or permanent disability. *Income protection* (briefly *IP*) policies in particular pay a periodic (usually weekly or monthly) income to an individual if he/she is prevented by sickness or injury from working and hence from getting the usual income. In the event of permanent disability, the benefit can consist of a lump sum (instead of a sequence of periodic amounts).

Disability benefits can be paid by individual disability insurance, group insurance or pension plans. In the first case, the disability cover may be a stand-alone cover or it may constitute a rider benefit in a life insurance policy, such as an endowment policy or a Universal Life product. A more detailed description of benefits provided by disability insurance is given in the following sections.

Various names are actually used to denote disability insurance products, in particular providing income protection, and can be taken as synonyms. The following list is rather comprehensive: *disability insurance*, *permanent health insurance* (the old British name), *income protection insurance* (the present British name), *loss-of-income insurance*, *loss-of-time insurance* (often used in the US), *long-term sickness insurance*, *disability income insurance*, *permanent sickness insurance*, *non-cancelable sickness insurance*, *long-term health insurance* (the last two terms are frequently used in Sweden).

3.5.1 Type of Benefits in Individual Policies

In individual disability insurance policies, permanent or not-necessarily permanent disability is considered according to the product design. Moreover, some disability policies only allow for total disability, whereas other policies also allow for partial disability. We describe the most important types of disability benefits.

The *disability income benefit* is a fixed-amount annuity benefit providing income protection in the case of total disability.

Various definitions of total disability are used. Some examples are as follows:

- the insured is unable to engage in his/her own occupation;
- the insured is unable to engage in his/her own occupation or carry out another activity consistent with his/her training and experience;
- the insured is unable to engage in any gainful occupation.

When one of the above definitions is met to a certain degree only, *partial disability* occurs.

Some policies provide a *lump sum benefit* in the case of permanent (and total) disability. The cover may be a stand-alone cover or it may be a rider to a basic life insurance policy, say an endowment insurance. It must be pointed out that moral hazard is present in this type of product design, which involves the payment of a lump sum to an individual who may subsequently recover (partially or fully), so that the benefit is irrecoverable in this event. A qualification period is commonly applied to limit the moral hazard.

The *waiver-of-premium benefit* is a rider benefit in a basic life insurance policy (e.g. an endowment, a whole-life assurance, etc.). The benefit consists of the waiver of life insurance premiums during periods of disability.

3.5.2 Types of Benefits in Group Insurance and Pension Plans

Disability group insurance may represent an important part of an employee benefit package (see also Sect. 3.10). There are two main types of benefits provided by disability group insurance:

- the *short-term disability (STD) benefit* protects against loss of income during short disability spells;
- the *long-term disability (LTD) benefit* protects against long-term (and possibly permanent or lasting to retirement age) disabilities.

The two fundamental types of disability benefits which can be included in a pension plan are as follows:

- a benefit consisting of an *annuity to a disabled employee*;
- a benefit consisting of a *deferred annuity to a (permanently) disabled employee*, beginning at retirement age.

The second type of benefit is usually found when a LTD group insurance operates (outside the pension scheme), providing disability benefits up to the retirement age.

3.5.3 Benefit Amount and Policy Conditions

Disability insurance should be distinguished from other products, within the area of health insurance. In particular, (short-term) sickness insurance usually provides reimbursement of medical expenses and hospitalization benefits, i.e. a daily benefit during hospital stays (see Sect. 3.4). Long-term care insurance provides income support for the insured, who needs nursing and/or medical care because of chronic or long-lasting conditions or ailments (see Sect. 3.6). A Critical illness (or Dread disease) policy provides the policyholder with a lump sum in case of a dread disease, i.e. when he/she is diagnosed as having a serious illness included in a set of diseases specified by the policy conditions; the benefit is paid on diagnosis of a specified condition, rather than on disablement (see Sect. 3.7). Note that in all these products the payment of benefits is not directly related to a loss of income suffered by the insured, whereas a strict relation between benefit payment and working inability characterizes the disability annuity products providing income protection.

In this chapter (as well as in Chap. 6) we mainly focus on individual policies providing disability annuities. Nevertheless, a number of definitions in respect of individual disability products also apply to group insurance and pension plans as well.

In individual disability insurance the size of the insured benefit needs to be carefully considered by the underwriter at the time of application, in order to limit the effects of moral hazard. In particular, the applicant's current earnings and the amount of benefits expected from other sources (social security, pension plans, etc.) in the event of disablement must be considered. When the insurance policy also allows for partial disability, the amount of the benefit is scaled according to the degree of disability, hence a graded benefit is paid.

In disability group insurance, benefits while paid are related to pre-disability earnings, typically being equal to a defined percentage (e.g. 70 %) of the salary. In pension plans the benefit payable upon disability is commonly calculated using a given benefit formula, normally related to the formula for the basic benefit provided by the pension plan.

Some disability covers pay a benefit of a constant amount while others provide a benefit which varies in some way. In particular, the benefit may increase in order to (partially) protect the policyholder from the effects of inflation. There are various methods by which increases in benefits are determined and financed. In any case, policies are usually designed so that increasing benefits are matched by increasing premiums. Benefits and premiums are often linked to some index, say an inflation rate, in the context of an *indexation mechanism*.

Another feature in disability policy design consists in the *decreasing annuity benefit*. In this case, the benefit amount reduces as the past duration of the disability claim increases. Such a mechanism is designed in order to encourage a return to gainful work.

Individual disability policies include a number of conditions, in particular concerning the payment of the insured benefits. Conditions which aim at defining the time interval during which benefits can be paid have been described in Sect. 3.2.4. These policy conditions have a special importance from an actuarial point of view, when calculating premiums and reserves. Examples will be provided in Sects. 6.7 and 6.11.

3.6 Long-Term Care Insurance

Long-term care insurance (LTCI) provides the insured with financial support, while he/she needs nursing and/or medical care because of chronic (or long-lasting) conditions or ailments.

Several types of benefits can be provided (fixed-amount annuities, care expense reimbursement, etc.; see Sect. 3.6.2). The benefit trigger is usually given either by claiming for nursing and/or medical assistance (together with a sanitary ascertainment), or by assessment of individual disability, according to some predefined metrics (e.g. the ADL method, see the next section).

Remark LTCI products deserve an extensive presentation for various reasons. On the one hand, LTCI provides benefits of remarkable interest in the current demographic and social context. On the other hand, LTCI covers are "difficult" insurance products both from the insurer's and the individual perspective. We note, in particular, the following aspects.

- In many countries, the elderly population is rapidly growing because of increasing life expectancy and low fertility rates.
- Household size is progressively reducing, with a consequent lack of assistance and care services provided to old members of the family inside the family itself.
- LTCI products are rather recent (especially if compared to other insurance products in the framework of health insurance, e.g. personal accident insurance and sickness insurance). As a consequence, senescent disability data are scanty, so that pricing difficulties arise.
- High premiums (for instance, because of a significant safety loading) can be an obstacle to the diffusion of these products. Further, it should be stressed that, in its stand-alone format, the LTCI product only provides "protection", excluding other types of benefits (e.g. a straight life annuity or a death benefit). As we will see in Sect. 3.6.3, appropriate packaging of LTCI benefits together with lifetime-related benefits can enhance the clients' propensity to purchase LTCI products. ■

3.6.1 Measuring the Severity of Disability

According to the *ADL* (*Activities of Daily Living*) *method*, the activities and functions considered are, for example, the following:

1. eating
2. bathing
3. dressing
4. moving around
5. personal hygiene
6. going to the toilet

The simplest implementation of the ADL method is as follows. For each activity or function, the individual ability is tested. The total disability level (or *LTC score*) is given by the number of activities or functions the insured is not able to perform, and, finally, it is expressed in terms of *LTC state*. See Table 3.2, where an example of graded LTCI benefit is also given.

More complex implementations of the ADL method rely on the degrees of ability to perform the various activities (see Example 3.6.1).

The *IADL (Instrumental Activities of Daily Living) method*, also known as the *PADL (Performance Activities of Daily Living) method*, is based on the individual ability to perform "relation" activities; for example: ability to use a telephone, shopping, food preparation, housekeeping, etc.

The Barthel index and the OPCS index constitute two important examples of methods for assessing the disability severity, that is, the level of functional dependence. The interested reader can refer to Sect. 3.13 for bibliographical suggestions. As regards the OPCS index, see Example 3.6.1.

Example 3.6.1 The *OPCS index* is based on the degree of functional dependence in performing 13 activities (among which are mobility, eating, drinking, etc.). The index quantifying the overall disability of a generic individual is calculated according to the following procedure:

1. the degree p_j is assessed for each activity j, $j = 1, 2, \ldots, 13$;
2. let $p^{(1)}$, $p^{(2)}$, $p^{(3)}$ denote the three highest values among the p_j's ($p^{(1)} \geq p^{(2)} \geq p^{(3)}$);
3. the overall degree, p, is determined as follows, that is, via a weighting formula:

$$p = p^{(1)} + 0.4\, p^{(2)} + 0.3\, p^{(3)}; \tag{3.6.1}$$

4. the value of p determines the "category" and the "level" of disability (which are also used in various statistical reports); see Table 3.3. □

Table 3.2 Benefit as a function of the LTC state

ADL score; unable to perform:	LTC state	Graded benefit (% of the insured benefit)
3 Activities	I	40
4 or 5 Activities	II	70
6 Activities	III	100

Table 3.3 Disability categories and levels according to OPCS index

p	Category	Level
0.5–2.95	1	–
3.0–4.95	2	–
5.0–6.95	3	–
7.0–8.95	4	–
9.0–10.95	5	–
11.0–12.95	6	I
13.0–14.95	7	I
15.0–16.95	8	I
17.0–18.95	9	II
19.0–21.40	10	II

Remark It has been stressed that a weak point of disability assessment via ADL (or IADL) can be found in possible significant correlations among an individual's ability to perform the various activities. The consequence is a likely concentration of insureds in the "extreme" categories, i.e. those with either a very low or a very high disability degree. ■

3.6.2 LTCI Products: A Classification

Long-term care insurance products can be classified as follows:

- products which pay out benefits with a *predefined amount* (usually, a lifelong annuity benefit); in particular
  - a *fixed-amount* benefit;
  - a *degree-related* (or *graded*) benefit, i.e. a benefit whose amount is graded according to the degree of disability, that is, the severity of the disability itself (for example, see Table 3.2);
- products which provide reimbursement (usually partial) of nursery and medical expenses, i.e. *expense-related* benefits;
- *care service* benefits (for example, provided by the Continuing Care Retirement Communities, briefly CCRCs); see Sect. 3.6.5.

3.6.3 Fixed-Amount and Degree-Related Benefits

A classification of LTCI products which pay out benefits with a predefined amount is proposed in Fig. 3.7.

Immediate care plans, or *care annuities*, relate to individuals already affected by severe disability (that is, in "point of need"), and then consist of:

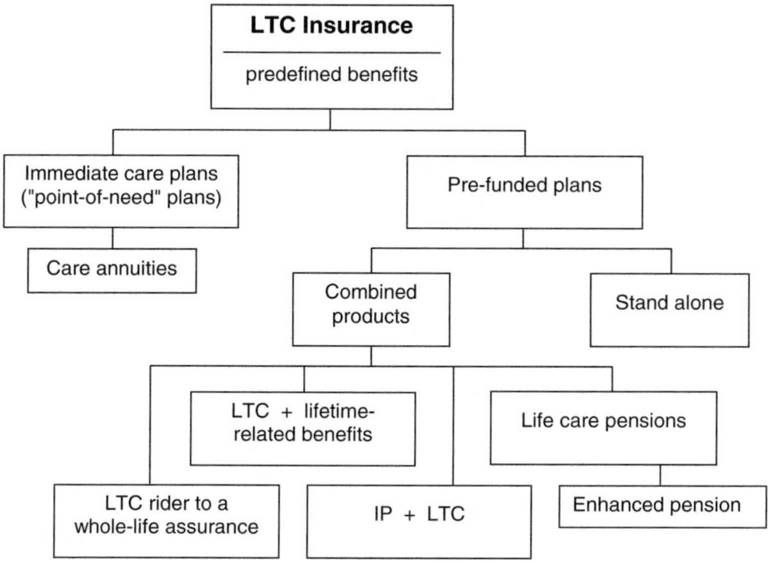

Fig. 3.7 A classification of LTC insurance products providing predefined benefits

- the payment of a single premium;
- an immediate life annuity, whose annual benefit may be graded according to the disability severity.

Hence, care annuities are aimed at seriously impaired individuals, in particular persons who have already started to incur long-term care costs. The premium calculation is based on assumptions of short life expectancy. However, the insurer may limit the individual longevity risk by offering a limited term annuity, i.e. a temporary life annuity.

Remark Care annuities belong to the class of *special-rate annuities*, also called *underwritten annuities*, because of the ascertainment of higher mortality assumptions via the underwriting requirements. The following special-rate annuities are sold in several markets.

1. The underwriting of a *lifestyle annuity* takes into account smoking and drinking habits, marital status, occupation, height and weight, blood pressure and cholesterol levels.
2. The *enhanced annuity* pays out an income to a person with a slightly reduced life expectancy, in particular because of a personal history of medical conditions. Of course, the "enhancement" in the annuity payment (compared to a standard life annuity, same premium) comes, in particular for this type of annuity, from the use of a higher mortality assumption.
3. The *impaired-life annuity* pays out a higher income than an enhanced annuity, as a result of medical conditions which significantly shorten the life expectancy of the annuitant (e.g. diabetes, chronic asthma, cancer, etc.).

4. Finally, *care annuities* are aimed at individuals, usually beyond age 75, with very serious impairments or individuals who are already in a LTC state.

Thus, moving from type 1 to type 4 results in progressively higher mortality assumptions, shorter life expectancy, and hence, for a given single premium amount, in higher annuity incomes. In particular, as regards annuities of types 3 and 4, the underwriting process must result in classifying the applicant as a *substandard risk*. ■

Pre-funded plans consist of:

- the accumulation phase, during which periodic premiums are paid; the accumulation can degenerate in a single premium;
- the payout period, during which LTC benefits (usually consisting of a life annuity) are paid in the case of LTC need.

Several products belong to the class of pre-funded plans. A *stand-alone LTC cover* provides an annuity benefit, possibly graded according to an ADL or IADL score. This cover can be financed by a single premium, by temporary periodic premiums, or lifelong periodic premiums. Of course, premiums are waived in the case of an LTC claim. This insurance product only provides a "risk cover", as there is, of course, no certainty in future LTC need and the consequent payment of benefits.

Two possible individual stories are considered in Fig. 3.8, one of which leads to the LTC benefit payment.

A number of *combined products* have been designed, mainly aiming at reducing the relative weight of the risk component by introducing a "saving" component, or by adding the LTC benefits to an insurance product with a significant saving component. Some examples follow.

LTC benefits can be added as a *rider to a whole-life assurance* policy. For example, a monthly benefit of, say, 2 % of the sum assured is paid in the case of an LTC claim, for 50 months at most. The death benefit is consequently reduced, and disappears if all the 50 monthly benefits are paid. Thus, the (temporary) LTC annuity benefit consists in an *acceleration* of the death benefit. The LTC cover can be complemented by an additional deferred LTC annuity (financed by an appropriate premium increase)

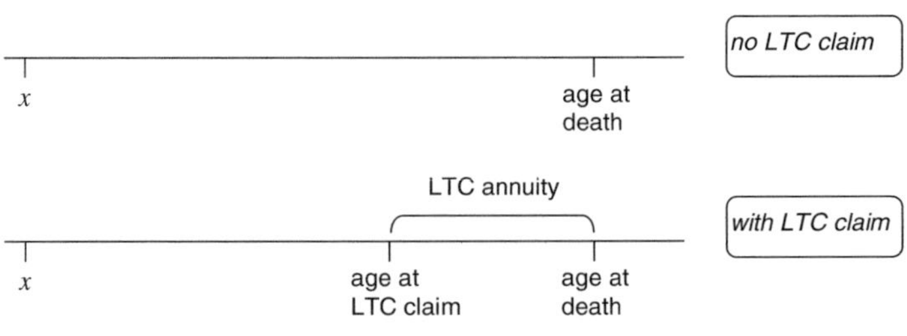

Fig. 3.8 LTC stand-alone annuity benefit: possible outcomes

which will start immediately after the possible exhaustion of the sum assured (that is, if the LTC claim lasts for more than 50 months) and will terminate at the insured's death.

Three possible individual stories are considered in Fig. 3.9, namely: no LTC claim, "long" LTC claim (implying exhaustion of the sum assured), "short" LTC claim. In the case of a long LTC claim, the possibility of an additional LTC annuity is also considered.

An insurance package can include LTC benefits combined with *lifetime-related benefits*, i.e. benefits only depending on insured's survival and death; more precisely:

1. a lifelong LTC annuity (from the LTC claim on);
2. a deferred life annuity (e.g. from age 80), while the insured is not in LTC disability state;
3. a lump sum benefit on death, which can alternatively be given by:
    a. a fixed amount, stated in the policy;
    b. the difference (if positive) between a stated amount and the amount paid as benefit 1 and/or benefit 2.

Four possible individual stories and the consequent outcomes in terms of benefits are shown in Fig. 3.10.

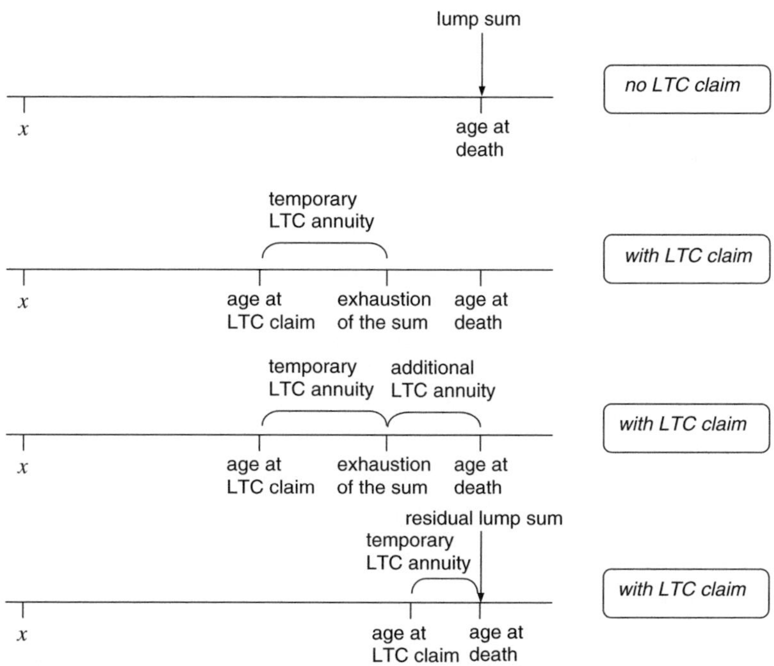

Fig. 3.9 LTC (temporary) annuity as an acceleration benefit in a whole-life assurance: possible outcomes

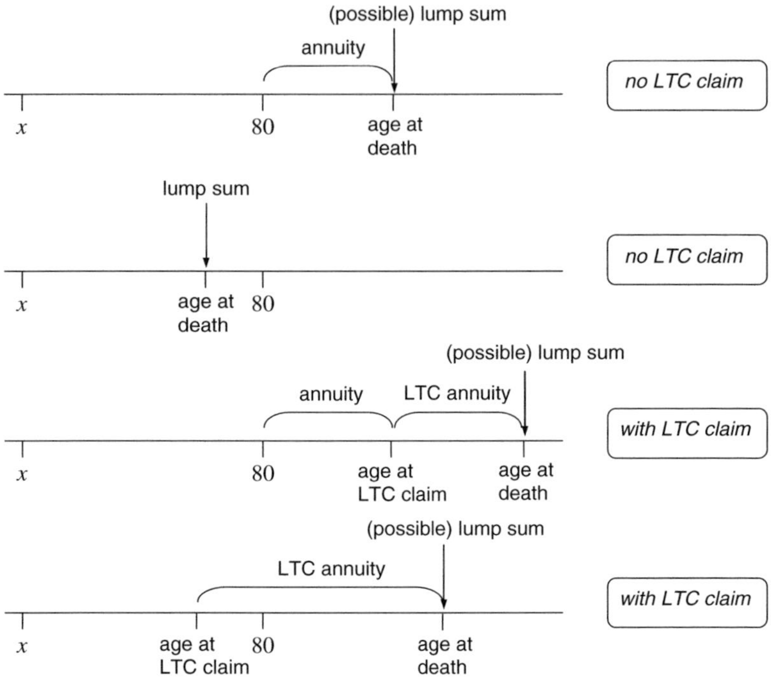

Fig. 3.10 Insurance package including LTC annuity and lifetime-related benefits: possible outcomes

Life care pensions (also called *life care annuities*) are life annuity products in which the LTC benefit is defined in terms of an uplift with respect to the basic pension. The basic pension b is paid out from retirement onwards, and is replaced by the LTC annuity benefit $b^{[\mathrm{LTC}]}$ ($b^{[\mathrm{LTC}]} > b$) in the case of an LTC claim. The uplift can be financed during the whole accumulation period by premiums higher than those needed to purchase the basic pension b.

The *enhanced pension* is a particular life care pension in which the uplift is financed by a reduction (with respect to the basic pension b) of the benefit paid while the policyholder is healthy. Thus, the reduced benefit $b^{[\mathrm{healthy}]}$ is paid out as long as the retiree is healthy, while the uplifted benefit $b^{[\mathrm{LTC}]}$ will be paid in the case of an LTC claim (of course, $b^{[\mathrm{healthy}]} < b < b^{[\mathrm{LTC}]}$).

Two possible individual stories and the consequent outcomes in terms of benefits are shown in Fig. 3.11, which can be referred to the life care pension scheme (according to which the basic pension benefit b is paid while the retiree is healthy) and the enhanced pension scheme (which implies, while the retiree is healthy, a reduced pension benefit $b^{[\mathrm{healthy}]}$).

The pension benefits and the LTC benefit are illustrated in Fig. 3.12 (a higher premium is implicitly assumed to finance the uplift) and Fig. 3.13 (which shows the reduction of the pension benefit to finance the uplift).

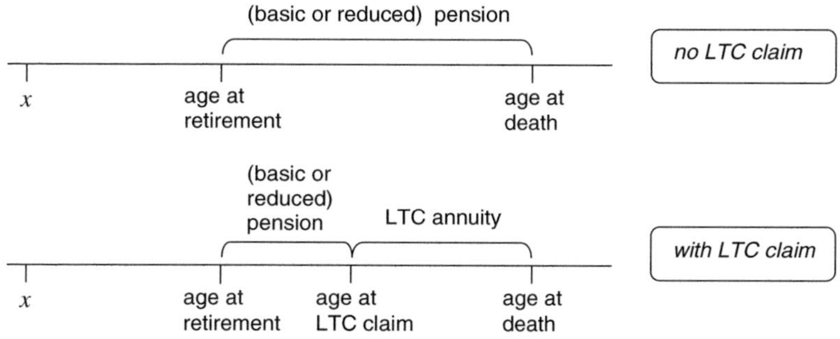

Fig. 3.11 Life care pension and enhanced pension: possible outcomes

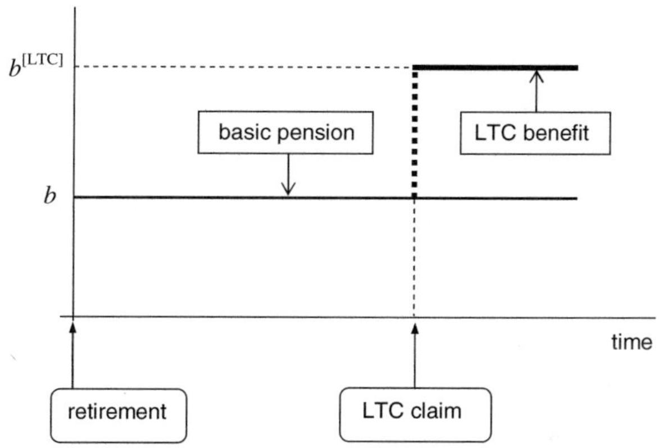

Fig. 3.12 Benefits provided by a life care pension product

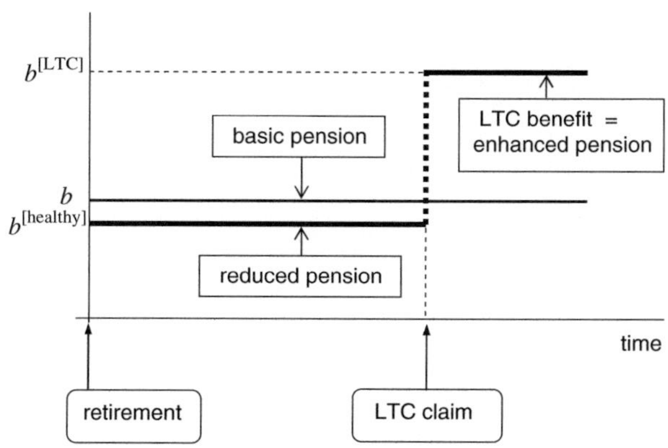

Fig. 3.13 Benefits provided by an enhanced pension product

Finally, a *lifelong disability cover* can include:

- an income protection cover (briefly, IP; see Sect. 3.5.1) during the working period, that is, during the accumulation period related to LTC benefits;
- an LTC cover during the retirement period.

3.6.4 Expense-Related Benefits

This category includes LTCI products which provide expense reimbursement. Two basic types of products can be recognized.

Stand-alone LTC cover, whose benefits consist in (partial) reimbursement of expenses related to LTC needs, in particular nursery, medical expenses, physiotherapy, etc. Usually, there are limitations on eligible expenses; further, deductibles as well as limit values are stated in the policy conditions.

LTC benefits can also be provided by an *LTC cover as a rider to sickness insurance*. The resulting product is a lifelong sickness insurance. In order to cover LTC needs, eligible expenses are extended, so to include, for example, nursing home expenses. Further, a fixed-amount daily benefit can be provided for expenses without documentary evidence.

3.6.5 Service Benefits

The LTCI products providing care service benefits usually rely on an agreement between an insurance company and an institution which acts as the care provider.

An interesting alternative is given by the Continuing Care Retirement Communities, briefly CCRCs, which have become established in the US. CCRCs offer housing and a range of other services, including long-term care. The cost is usually met by a combination of entrance charge plus periodic fees (that is, upfront premium plus monthly premiums).

3.7 Critical Illness Insurance

A Critical Illness Insurance (CII), or Dread Disease (DD), policy provides the policyholder with a lump sum in case of a severe illness, i.e. when he/she is diagnosed as having an illness included in a set of diseases specified by the policy conditions. The most commonly covered diseases are heart attack, coronary artery disease requiring surgery, cancer and stroke. However, CII products have a very limited extension of coverage, defined via listing (rather than via exclusions).

3.7.1 Types of Benefits

It is important to stress that the benefit is paid on diagnosis of a specified condition, rather than on disablement. Hence, a CII policy differs in its objectives from other policies discussed in this chapter. In particular, the CII cover:

- does not indemnify the insured against any specific loss due to medical expenses (whereas medical expense reimbursement is provided by sickness insurance; see Sect. 3.4.1);
- does not meet any specific income need, arising from loss of earnings (which is conversely met by an income protection policy; see Sect. 3.5.1).

Of course, the CII benefit can help in covering medical expenses and providing protection against possible loss of income.

Policies are usually for a fixed term (say, 5 or 10 years), and include a waiting period (starting at policy issue) so that the CII benefit will not be payed if the diagnosis falls in the waiting period itself; the waiting period aims at reducing possible adverse selection.

The following benefit arrangements can be recognized.

1. A *stand-alone* cover only includes a CII benefit; the insurance policy ceases immediately after the payment of the sum assured.
2. A CII benefit can constitute a *rider benefit* for a life insurance cover, in particular a term insurance providing a benefit in the case of death. A CII rider benefit takes either of the two following forms.
    - *Additional benefit*: According to this arrangement, the insurance policy includes two separate covers (possibly with different sum assured), one paying the sum assured in the case of death, and the other paying the sum assured in the case of critical illness. After the payment of the CII benefit, only the term insurance providing the death benefit remains in force.
    - *Acceleration benefit*: The arrangement is defined as follows. Let S denote the sum assured in the life insurance cover; the amount λS (with λ stated in the policy conditions, $0 < \lambda \leq 1$) is payable on critical illness diagnosis, while the remaining amount $(1-\lambda)S$ is payable on death, if this occurs within the policy term. Note that, if $\lambda = 1$ the whole insurance cover ceases after the payment of the CII benefit.

In the case of an additional benefit, it is important to avoid a possible situation of "overpayment" (which implies higher expected costs and hence higher premiums), that could take place when death occurs within a very short period after disease inception. A solution is achieved by replacing the single cash payment with a series of payments (for example, three or four semestral or annual payments), each payment being conditional on the survival of the insured.

It is reasonable to assume that the (periodic) premiums are payable while the insured is healthy, even if the CII benefit is a rider to a life insurance policy, so that the life cover does not cease at the critical illness diagnosis.

3.7.2 Multiple Critical Illness Benefits

The classical CII product, in its stand-alone version, terminates after the (first) claim and the payment of the related benefit. In other words, it can be labelled as a single-payment insurance product. However, the need for protection against further possible serious illnesses can last beyond the claim. Therefore, insurance products providing coverage extended to more than one critical illness claim can offer a more complete protection. We also note that, thanks to medical advances in recent decades, the life expectancy after some critical events has increased, so that the need for further protection extends over a period longer than in the past.

Two alternative approaches can be adopted to the construction of a CII product which provides possible multiple benefits.

1. Multiple CII benefits can be paid by a *buy-back CII product*, that is, a classical CII product with a "buy-back" option as a rider which gives the right to reinstate the CII cover after the first claim; the (second) CII cover is then sold without medical assessment and without change in the premium rates, after a waiting period (1 year, say) following the first claim. The option must be chosen at policy issue. Usually, the same or related type of illness is excluded from the second coverage.
2. A specific *multiple CII cover* can be designed (usually as a stand-alone cover) in order to provide multiple CII benefits. The "grouping approach" is usually adopted in order to classify the diseases and determine appropriate exclusions. In general, after a claim due to a disease belonging to a given group, all the diseases included in that group (and hence highly correlated) are excluded from further coverage.

3.8 Other Limited-Coverage Products

CII products are typical examples of limited-coverage health insurance products, as only diseases belonging to a well defined set are covered. Other limited-coverage products have been recently launched in some insurance markets. Here we focus on cancer insurance policies and surgery cash plans.

A *cancer insurance* policy can be shaped in several different ways, also depending on the specific insurance market. There are essentially two main types of benefit.

The *lump sum benefit* consists of a single payment upon the diagnosis of a cancer. This fixed-amount benefit can be used in any way, not necessarily related to the medical expenses the insured incurs (for example: ground and air transportation, private nursing, etc.).

The other type of benefit, usually called an *expense benefit plan*, consists of a set of payments, each payment being related to a specific expense item, typically: medical tests, hospital stay, surgery, radiation, chemotherapy, etc. The amount of each payment is predefined in the policy, so that this type of benefit can also be classified as a fixed-amount benefit, although the total amount paid-out depends on the specific needs.

Underwriting requirements are applied. To qualify, the applicant should not have any pre-existing conditions related to cancer.

A *surgery cash plan* provides the insured with a cash benefit if he/she requires any type of medically necessary in-patient or day surgery. A waiting period is commonly applied to avoid adverse selection, whereas no particular underwriting requirements are usually applied, at least for given age ranges at policy issue (in this case, a surgery cash plan is a guaranteed-issue product).

The basic benefit is a *lump sum benefit*. The amount paid out depends on the sum insured, and then varies according to the severity of the operation and the recovery period required. Hence, it can be considered as a graded benefit. We note that:

- the cash benefit is not a reimbursement benefit;
- the insured can use the cash amount for any purpose, including post-surgery care, physiotherapy, etc.

As a supplementary benefit, a fixed-amount daily benefit could be added to the policy. This benefit is payable during the hospital stay.

3.9 Combining Health and Life Benefits

It is worth noting that, from the insurer's perspective, a combined product can be profitable even if one of its components is not profitable. Further, from a specific risk management perspective, packaging several insurance covers into one policy leads to a total amount of policy reserve which can constitute a policy "cushion" for facing poor experience inherent in one of the package components, provided that some degree of flexibility in using available resources is allowed.

Conversely, from the client's perspective, purchasing a combined product can be less expensive than purchasing each single component, in particular thanks to a reduction of the acquisition costs charged to the policyholder.

3.9.1 Health Covers as Riders to Life Insurance

The simplest and most traditional way to combine benefits, in the framework of the insurances of the person, is to define a health-related benefit (or a cause-of-death-related benefit) as a rider to a life insurance policy (viz a term insurance, a whole-life insurance, an endowment insurance, etc.).

Several examples have already been mentioned in previous sections (see Sects. 3.3–3.7). Further details follow. The reader can also refer to Fig. 3.14 for some examples.

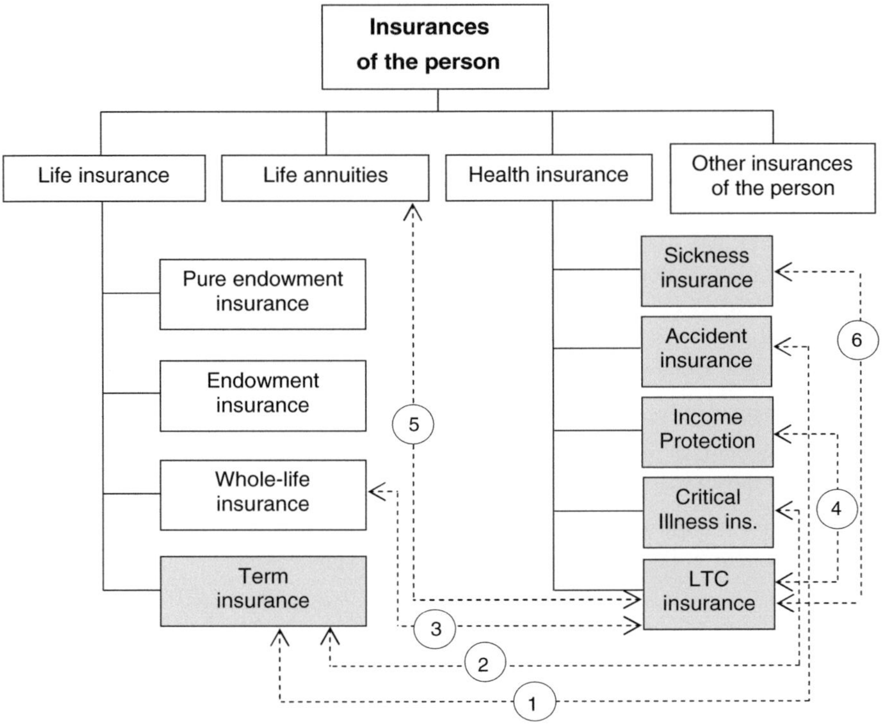

Fig. 3.14 Combining health and life benefits

Accident insurance benefits (see Sect. 3.3.1) can constitute riders to a life insurance policy which includes a death benefit, e.g. a term insurance (see link 1 in Fig. 3.14). In particular, the sum insured as the death benefit can be paid in the event of permanent disability. Another type of rider provides, in the case of accidental death, an amount higher than the sum insured as the (basic) death benefit.

Critical illness benefit (see Sect. 3.7.1) can be provided as a rider to a term insurance (see link 2 in Fig. 3.14); in this case the CII benefit is an acceleration benefit.

Waiver of premiums is a frequent rider benefit in several life insurance policies: premiums are waived in the event of (total) disability, over the whole disability spell (see Sect. 3.5.1).

3.9.2 Health Covers in Insurance Packages

Combining LTCI benefits with lifetime-related benefits leads to complex insurance packages (rather than simple rider benefits).

As seen in Sect. 3.6.3, various products integrate LTC guarantees into the saving process (as is the case for whole-life insurance, see link 3), or in the pension payout phase (see link 5).

Moreover, LTCI benefits can be packaged with other health-related benefits, for example with income protection (link 4), or with lifelong sickness insurance (link 6).

Universal Life (UL) policies are typical products in the US market, which can be designed either as participating or unit-linked policies. Their main features consist in a high flexibility available to the policyholder in deciding year by year: the amount of premium, to make a partial withdrawal, the type of investment backing the reserve, and so on. Further, similarly to a bank account, the policyholder receives a periodic statement, showing the costs (acquisition costs, management fees, fees for rider benefits, etc.) that have been charged to his/her policy account. If the policy is designed on a unit-linked basis, the current value of the fund is reported in the statement; if a participating arrangement is designed, the statement reports the annual adjustment which has been credited to the fund. The structure of a UL policy is shown in Fig. 3.15.

The underlying contractual form is a whole-life assurance. This way, the contract has no specified maturity; the contract terminates either because of death or full withdrawal. The death benefit is defined so that the sum at risk (that is, the difference between the death benefit and the fund) is positive.

Given the wide range of benefits which can be included, the UL policy can be regarded as an insurance package in the context of the insurances of the person. Beyond the death benefit, many health-related benefits can be included; for example: lump sum in the case of permanent disability, daily benefit in the case of temporary disability, medical expense reimbursement, etc. All these benefits (and the death benefit as well) can be financed, withdrawing the related annual (or periodic) cost from the fund, i.e. according to a natural premium-based arrangement (see Fig. 3.15).

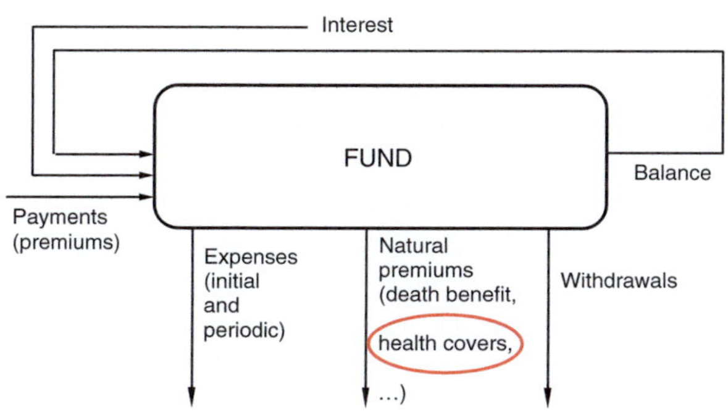

Fig. 3.15 Financing health insurance covers within a UL product

3.10 Group Insurance in the Health Area

Many health insurance products can be designed and sold on a group basis. The products are then referred to as *health group plans*, and provide coverage to a select group of people. The group typically consists of the employees of a firm, possibly extended to their dependents.

The usual benefit package first includes medical expense reimbursement (see Sect. 3.4.1), and in this case dependents may be included. Another component of the package may be income protection insurance (see Sect. 3.5.2).

Health group insurance provided to the employees of a firm may be either compulsory or voluntary. In the former case, all the employees are members of the health plan, whereas in the latter all eligible employees may decide to opt for the group cover. Adverse selection does not affect health plans with compulsory membership, and hence the underwriting requirements (if any) are very weak; conversely, underwriting requirements are appropriate in the case of voluntary membership.

Whatever the membership arrangement, moral hazard can be limited by adopting appropriate deductibles.

Premium calculation can be organized either on an individual basis or on a group basis. The latter arrangement is typically applied when the membership is compulsory and the health cover is sponsored by the employer. In this case, premiums are usually calculated on a one-year basis taking into account the current structure of the insured group.

Health group plans can be placed in the framework of *employee benefit plans*. An employer can provide its employees with benefits other than the salary, among which we can find the following insurance-related benefits:

- death benefits, paid to the employee's dependents in the event of death during the working period;
- pensions, i.e. post-retirement benefits;
- health insurance covers.

In traditional health group plans, the benefit package and the related limitations (exclusions, deductibles, etc.) are defined in the group insurance policy. An alternative structure, implemented in the US in particular, can be found in the *Defined Contribution Health Plans* (briefly, DCHPs).

Following the shift from "defined benefit pension plans" to "defined contribution pension plans", a DCHP relies on the same logic as regards the employer's contributions. Instead of paying premiums which depend on a defined package of health-related benefits, the employer pays a defined amount (that is, a contribution) to each employee. The employees can then purchase individual health policies on the insurance market, according to his/her needs and preferences.

A DCHP can be implemented in different ways. The structure described above implements the so-called "pure" DCHP, or "individual market model of DCHP". An alternative model is the "decision support model", according to which the employer's defined contributions fund for each employee a health-savings account (see Sect. 1.3) and a health insurance cover (usually with high deductibles) within a health group policy. This way, difficulties in the choice of individual policies are eliminated, while the employees retain, to some extent, the possibility of choosing an appropriate health insurance cover.

3.11 Public and Private Health Insurance

Health insurance products sold by (private) insurance companies constitute the main topic we are addressing. However, it is worth noting that a large variety of health insurance arrangements can be found in different countries. In particular, mixed systems of health care funding, which rely on both *public health insurance* and *private health insurance*, are rather common.

Public health insurance is mainly financed through income-related taxation or contributions, whereas private health insurance basically relies on insurance products financed through premiums whose amount depends on the value of the benefit package, e.g the expected present value calculated according to actuarial principles. Private insurance premiums can be calculated either on an individual basis or on a group basis (for example in health group insurance; see Sect. 3.10).

Various "interactions" between public and private health insurance can be observed in different countries, as a result of the local legislation. For instance, participation into the public health insurance scheme can be mandatory either for the whole population or for eligible groups only, while it may be voluntary for specific population groups. Private health insurance is voluntary in most countries, while a basic health coverage is mandatory in some countries.

As private health insurance is the object of this book, it is interesting to focus on a classification of the main functions of health insurance products.[1]

1. *Primary private health insurance* is the health insurance that represents the only available access to basic coverage for individuals who do not have public health insurance, either because there is no public health insurance, or because individuals are not eligible to coverage under public health insurance (see (a) below), or they are entitled for public coverage but have chosen to opt out of such coverage (see (b)); in particular:

   (a) *principal* private insurance represents the only available access to health coverage for individuals where a public insurance scheme does not apply;
   (b) *substitute* private insurance replaces health coverage which would otherwise be available from a public insurance scheme.

2. Private insurance can offer *duplicate covers*, i.e. coverage for health services which are already provided by public health insurance. Duplicate covers also

[1] The following classification has been proposed by OECD (2004b).

offer access to different providers or levels of service; it does not exempt individuals from contributing to public health insurance.

3. *Complementary covers* complement coverage of publicly insured services or services within principal/substitute health insurance (which pays only a proportion of qualifying care costs) by covering all or part of the residual costs not otherwise reimbursed.
4. *Supplementary covers* provide coverage for additional health services not covered by the public insurance scheme. Its extension depends on the local public health legislation, and may then include luxury care, long-term care, dental care, rehabilitation, alternative medicine, etc.

3.12 Microinsurance in the Health Area

A basic concept, which helps in understanding the meaning and the role of microinsurance, is that of "excluded population", that is, a population without participation or with inadequate participation in social life, or without a place in the consumer society. Examples are given, in several countries, by people active in the informal economy in urban areas and most of the households in rural areas, employees in small workplaces, self-employed and migrant workers. Exposure to accident and illness risks may be particularly significant among those people.

In general, socio-economic inequalities within populations of the same country can imply significant differences in health conditions. Exclusion from existing health insurance schemes, either public or private, has, as likely consequences, a more severe impact of morbidity, a higher mortality, and a lower life expectancy.[2]

The basic problem is then making health insurance widely accessible. How to provide health insurance is a government choice. In most cases the choice has been to rely on the insurance market to extend the health insurance coverage to individuals with no access to existent public insurance schemes, or to provide the coverage in the case of non-existent public insurance scheme. A discussion on the ability of private insurance to fill the gap is beyond the scope of this section (the interested reader can refer to the papers and technical reports cited in Sect. 3.13). We focus on some basic aspects of health microinsurance arrangements.

Remark The term "micro" can be interpreted in several ways. In particular, it can denote the limited extension of the covered population, which constitute only a part of a national population. An alternative interpretation refers to the small amount of each financial transaction generated within a microinsurance arrangement, mostly because of the low income of people who constitute the target of microinsurance initiatives. ■

[2] This approach to health microinsurance has been suggested by Dror and Jacquier (1999); see also the references therein.

Health microinsurance can provide people with coverage of several health-related risks, in particular: illness and possible consequent hospitalization, injury and possible related disability. Further, coverage can be extended to the financial consequences of early death (although the death risk can constitute an object of life microinsurance).

Several parties are involved in the implementation of a microinsurance programme, and different models can be recognized. In what follows we briefly describe some basic models that underly existing health microinsurance programmes.

Various models (denoted, in what follows, as arrangements 1 to 3) rely on a *health microinsurance scheme* (briefly HMIS), i.e. an institution which provides insurance covers to individuals. The range of tasks assumed by the HMIS and hence the degree of its involvement in the delivery and management of the insurance covers, as well as the role of other possible parties, vary according to the arrangement adopted for implementing the health microinsurance programme.

The ultimate target in most microinsurance arrangements is the *unit*, which consists of individuals sharing a common activity and/or living in a well-defined geographic area. A microinsurance arrangement can involve several units, so that an improvement in the diversification via pooling can be gained.

We first focus on the following arrangements and the relevant parties involved.

1. This arrangement relies on a partnership which involves, besides the HMIS, an insurance company and an institution acting as the health provider.

   - The HMIS is responsible for:
     - the marketing of the health insurance products (see below);
     - the delivery of the products to the clients in the units.
   - The insurance company is responsible for:
     - the design of the insurance products (although the appropriate types of products should be suggested by the HMIS);
     - the management of risks transferred by the individuals belonging to the units.
   - The health care provider delivers services such as hospitalization, surgery, etc.

   See Fig. 3.16. The weak point of this arrangement is the limited involvement of the HMIS in the development of the products, which are ultimately shaped by the insurance company. Conversely, the transfer of risks to the insurer constitutes an advantage for the HMIS.

2. The partnership only involves (besides the HMIS) a health care provider; then:

   - The HMIS is responsible for:
     - the design of the health insurance products;
     - the marketing of the products;
     - the delivery of the products to the clients in the units;
     - the management of the pool of risks.
   - The health care provider delivers services such as hospitalization, surgery, etc.

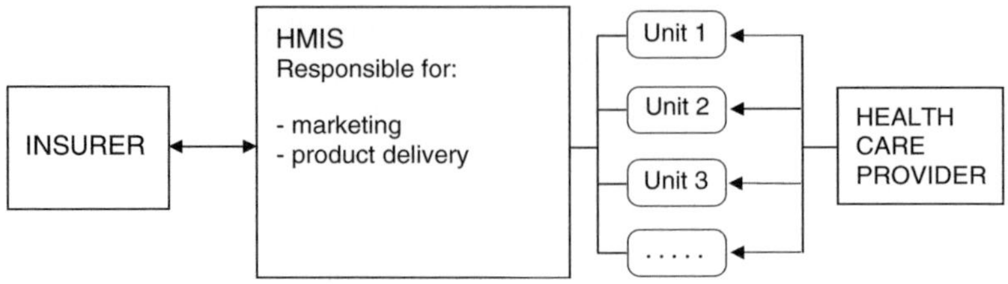

Fig. 3.16 HMIS-based health microinsurance arrangement (1)

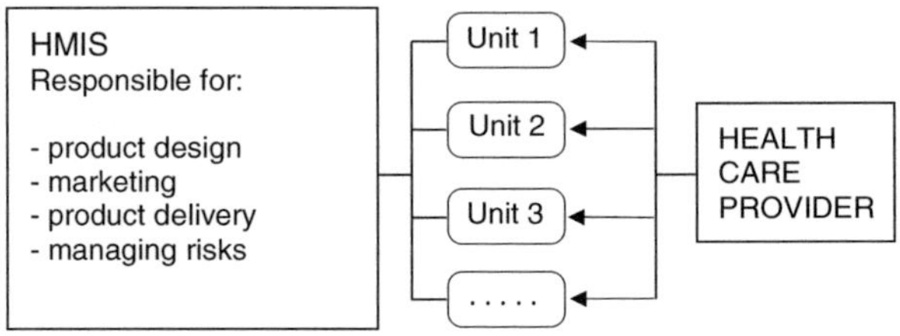

Fig. 3.17 HMIS-based health microinsurance arrangement (2)

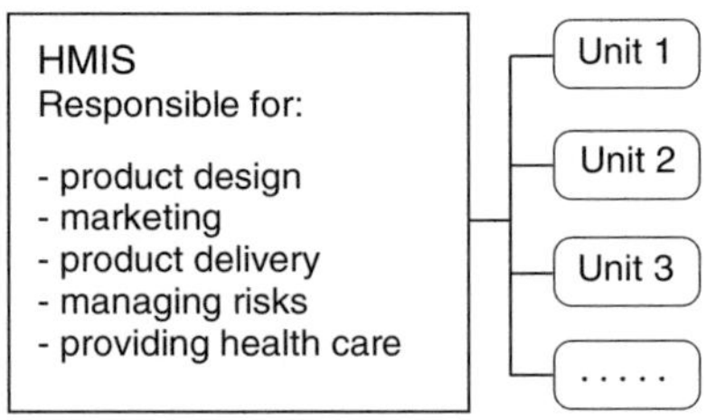

Fig. 3.18 HMIS-based health microinsurance arrangement (3)

See Fig. 3.17. The stronger involvement of the HMIS is a point in favor of this arrangement. A disadvantage is given by the risk taken by the HMIS as a consequence of the pool management.

3. The arrangement relies on the HMIS only, which also acts as the health care provider (being, at the same time, responsible for all the operations listed under arrangement 2). See Fig. 3.18. According to this arrangement, the HMIS has complete control of all the phases of the microinsurance process, of course bearing all the relevant risks.

A different approach to the implementation of a health microinsurance programme is adopted in the following arrangement.

4. This arrangement relies on a *mutual organization*. The individuals who constitute the community are, at the same time, insured and involved in all the operations. An external institution acts as the health care provider. See Fig. 3.19. While an

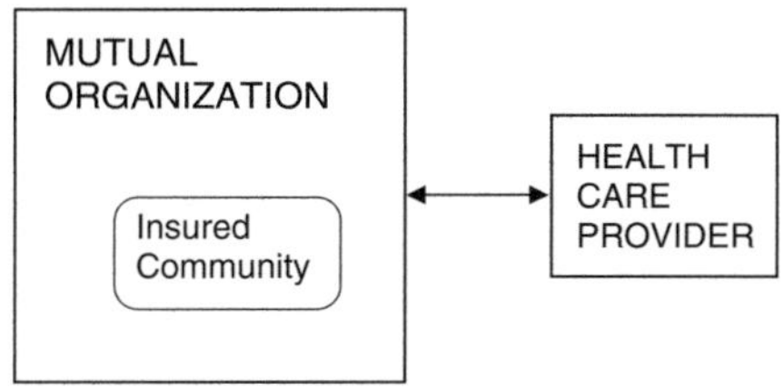

Fig. 3.19 Community-based health microinsurance arrangement

advantage is given by the full control on all the relevant operations, a weak point might be the possibly small size of the community and hence the poor diversification effect via risk pooling.

Finally, we note that any health microinsurance programme obviously implies an insurance activity, and can then be affected by all the problems concerning ordinary insurance. An example is given by the adverse selection effect: joining a microinsurance programme is usually not mandatory, and hence people heavily exposed to health-related risks might predominate in the membership.

3.13 Suggestions for Further Reading

Remark As already noted in Sect. 3.1, in the health insurance terminology the meaning of several terms is not univocally accepted. It follows that terms like "health insurance", "sickness insurance", "disability insurance" can have different meanings, which, to a large extent, depend on local traditions and market practice. This fact also reflects on the titles of books, papers, technical reports, etc. The following references are obviously chosen looking at the contents of the bibliographic material, disregarding to some extent the specific titles. ■

We first cite papers, books and technical reports which focus on (private) health insurance products. We note that, in several cases, actuarial issues are also addressed; these bibliographic items will also be cited in other chapters, in particular in Sects. 5.6 and 6.18.

Black and Skipper (2000), Bartleson (1968) and O'Grady (1988) deal with health insurance products in general, ranging from accident insurance to sickness insurance and disability annuities. A classification of health insurance products is proposed in OECD (2004b). Conversely, the following papers, reports and books refer to specific classes of health insurance products.

- Accident insurance: Alsina et al. (2003), and Jean (2004).
- Sickness insurance: Bernet and Getzen (2004), Fürhaupter and Brechtmann (2002), Milbrodt (2005), Newsom and Fernandez (2010), Orros and Webber (1988), and Szuch (2004).
- Disability insurance (in particular income protection): CMI (2006), Gregorius (1993), Haberman and Pitacco (1999), Mackay (1993), Pitacco (2004a), Sanders and Silby (1988), and Zayatz (2005).

- Long-term care insurance (LTCI): American Academy of Actuaries (1999), Dullaway and Elliott (1998), Gatenby (1991), and Jones (2004); in particular,
  - LTCI as a specific arrangement for the payout phase in life annuities and pension products: Brown and Warshawsky (2013), Murtaugh et al. (2001), Pitacco (2013), Warshawsky (2007), and Zhou-Richter and Gründl (2011);
  - Specific disability annuities: Ainslie (2000), and Rickayzen (2007);
  - Disability among the elderly and severity assessment (e.g. via ADL or IADL): Lawton and Brody (1969), Lafortune and Balestat (2004), Martin and Elliot (1992), and McDowell (2006).
  - Continuing Care Retirement Communities (LTC service benefits): Brecht et al. (2009), Jones (1995, 1996, 1997a,b), and Winklevoss and Powell (1984).
  - The economics and finance of LTC: Chen (1994), Costa-Font and Courbage (2012), Elliott et al. (2014), Gleckman (2010), Merlis (2004), Nuttall et al. (1994), Whynes (1996), and Zweifel (1996).
- Critical Illness Insurance (CII) (or Dread Disease insurance (DD)): de Braaf (2013), Brink (2010), Dash and Grimshaw (1993), Fabrizio and Gratton (1994), and Werth and Mannion (1998).

An overview on underwritten annuities (or special-rate annuities) is provided by Ridsdale (2012), specifically addressing the UK market. For information about other markets, see references therein.

Group insurance (and, in particular, health group insurance) is dealt with by Bluhm (1992), while Christianson et al. (2002) and Cucuta (2002) focus on Defined Contribution Health Plans.

As mentioned in Sect. 3.11, a large variety of health insurance arrangements can be found by looking at different countries, among which "mixed systems", relying on both public health insurance and private health insurance, are rather common. The literature devoted to the economics and finance of health insurance systems is very extensive. Cichon et al. (1999) provides a comprehensive view on financing public and private health care systems, and related modelling problems. Health care systems in European countries are analyzed in Jakubowski et al. (1998), Thomson and Mossialos (2009), Wendt (2009) and Wismar et al. (2011), whereas Colombo and Tapay (2004) and OECD (2004a, 2013) focus on OECD countries.

For more information on Health Microinsurance, the reader can refer to the following papers and reports: Dror and Jacquier (1999), Kimball et al. (2013), Leatherman et al. (2010), Le Roy and Holtz (2011), and Pott and Holtz (2013).

Chapter 4
Introduction to Actuarial Aspects

4.1 Some Preliminary Ideas

Actuarial aspects of health insurance modelling, for premium and reserve calculations, are strictly related to:

- types of benefits, in particular as regards their definition in quantitative terms (fixed-amount, degree-related amount, expense reimbursement; see Sect. 3.2.3);
- policy term (one-year covers versus multi-year covers, and possibly lifelong covers);
- premium arrangement (single premium, natural premiums, level premiums, etc.).

The premiums paid by the policyholders have to meet the benefits paid by the insurer, according to a stated criterion. We now assume that the premium is paid at policy issue (thus, no splitting into a sequence of periodic premiums is allowed for), and hence just one amount, namely a *single premium*, facing future benefits has to be determined.

Although the insurance business is based on the management of pools of risks, we approach premium calculation on an individual basis, namely referring to a single insured and the related insurance cover. Even though this approach might seem incomplete, as it does not explicitly allow for pooling effects, it is simple, and anyhow of great practical importance.

The (individual) premium must rely on some "summary" of the random benefits which will be paid by the insurer. Thus, in some sense, the premium represents a *value* of the benefits. As the benefits can consist, in general, of a sequence of random amounts paid throughout the policy duration, we have to summarize:

1. with respect to time, determining the random present value of the benefits, referred at the time of policy issue;
2. with respect to randomness, calculating some typical values of the probability distribution of the random present value of the benefits, namely the expected value, the standard deviation, and so on.

E. Pitacco, *Health Insurance*, EAA Series,
DOI 10.1007/978-3-319-12235-9_4

Step 1 requires the choice of the annual *interest rate* (or, more generally, the term structure of interest rates) for discounting benefits. It should be noted, however, that, when the policy duration is short (say, one year or even less), we can skip this step as the time does not have a significant impact on the value of benefits.

Step 2 first requires appropriate *statistical bases* in order to construct the probability distribution of the random present value of the benefits, and then the choice of typical values summarizing the distribution itself. The complexity of the statistical bases also depends on the set of individual *risk factors* accounted for in assessing the benefits (e.g. age, gender, health status, etc.), and the set of *rating factors*, among the risk factors, which are taken into account in the premium calculation.

So far we have only allowed for insurer's costs consisting in the payment of benefits. However, the insurer also has to pay *expenses* which are not directly connected with the amounts of benefits, for example general expenses. It is common practice to charge a share of these expenses to each insurance policy, via a convenient premium increase, that is, the *expense loading*.

Finally, a further increase in the premium amount provides the insurer with a *profit margin*. More details on loading for expenses and profit will be provided in Sect. 5.2.2.

The items listed above (i.e. statistical basis, interest rate, share of insurer's expenses, profit margin) constitute the ingredients of a "recipe", called the *premium calculation principle*, whose result is the *actuarial premium*. It is worth stressing the meaning of "actuarial". The output of the procedure described above is the premium calculated according to sound actuarial (i.e. financial and statistical) principles. Nonetheless, ingredients other than those so far considered can affect the actual *price* of the insurance product. For example, competition on the insurance market could suggest lowering the price in order to launch a more appealing product. In the following, we will not take these aspects into account.

Figure 4.1 summarizes the process leading to the price of an insurance product.

4.2 Technical Features of Premium Calculation

Pricing health insurance products relies, to a large extent, on a mixture of non-life insurance (or general insurance) and life insurance actuarial methods. This section points out some basic technical ingredients of both the non-life and the life actuarial fields.

Non-life and life technical features are summarized in Fig. 4.2, according to the type of benefit, the policy term and the premium arrangement.

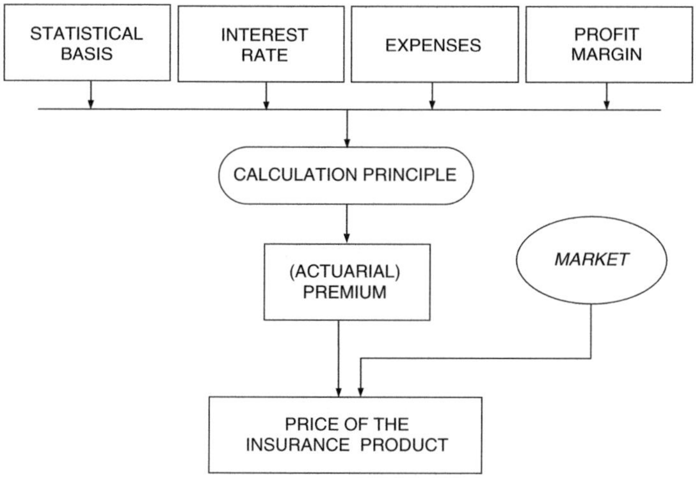

Fig. 4.1 Pricing an insurance product

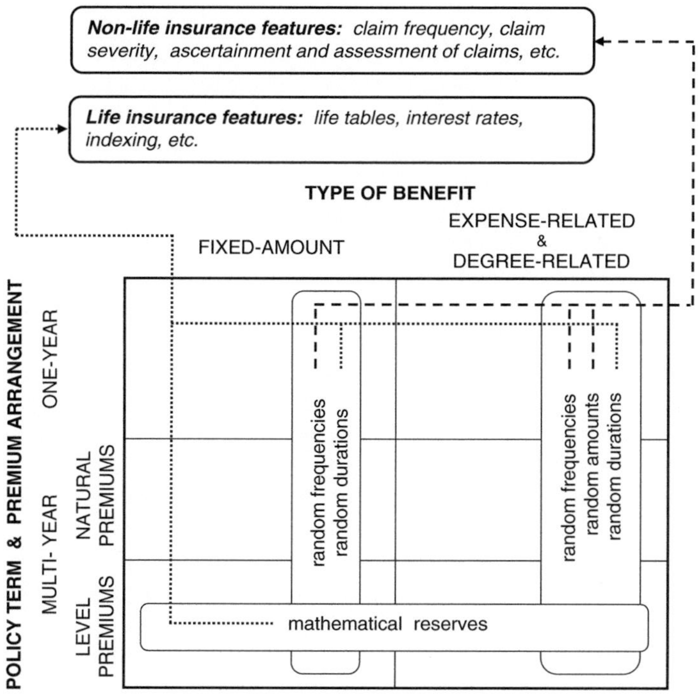

Fig. 4.2 Actuarial features of health insurance modelling

4.2.1 Non-life Insurance Aspects

We refer to the calculation structure described in Sect. 4.1 (see Fig. 4.1 in particular), and focus on statistical bases and expense issues.

The premium calculation for all the types of health insurance covers relies on the claim frequency, whereas the claim size only concerns insurance covers providing expense reimbursement, and covers in which the benefit is graded, i.e. it depends on some degree, for instance the degree of disability.

The expected *claim frequency* enters into the premium (and reserve) calculations according to the actuarial structure adopted. As we will see in the following chapters, sickness insurance products (dealt with in Chap. 5) significantly differ from disability insurance covers (see Chap. 6) as regards the expression of the expected claim frequency.

Sickness insurance models mainly rely on an expected claim frequency expressed as the expected number of claims per insured in a given coverage period (typically a year), while disability insurance covers providing annuity benefits require more complex models, allowing for the evolution of the individual health status throughout time. It follows that expected claim frequencies should be expressed in terms of the probability of entering into the disability state (or into one of the possible disability states). Nevertheless, in the context of practical approaches to disability annuity modelling, we also find a large variety of approximate calculation methods, most of which have been defined to comply with scanty or inappropriate statistical data; this aspect will be analyzed in Sects. 6.10–6.14.

Experience monitoring can help in improving the reliability of statistical data used in premiums (and reserves) calculations. For this purpose, experience rating procedures, typically relying on credibility theory results, should be adopted.

The *claim size* constitutes the other component of the statistical basis which underpins the premium calculation. A typical example of claim size is given by the amount of health-related expenses. While rigorous actuarial models should rely on the probability distribution of the claim size (as well as on the probability distribution of the claim number per policy), practical actuarial approaches usually account for the expected claim size only (an example will be provided in Sect. 5.2.1). Of course, the expected claim size is strictly related to policy conditions as the deductible and the limit value.

Estimation of claim sizes is also required for claim reserving purposes. However, it should be noted that, in health insurance, claim reserving has a lower importance than in other non-life insurance areas. Nevertheless, specific claim reserving problems may arise from the presence of a deferred period in disability annuity benefits; this aspect will briefly be addressed in Sect. 6.6.

As regards insurer's *expenses*, we note that, beyond shares of general expenses which are charged to the premium as is usual for all the insurance products, specific expenses related to the claim management must be accounted for in health insurance products. These expenses relate to the claim settlement (ascertainment and assessment of claims), as well as to the ascertainment of the claim prosecution throughout time, e.g. in insurance covers providing an annuity benefit in the case of not necessarily permanent disability.

4.2.2 Life Insurance Aspects

Life insurance actuarial aspects mainly refer to medium-term and long-term (possibly lifelong) products, in particular: disability insurance, long-term care insurance (LTCI) covers, and multi-year sickness insurance covers.

Again, refer to the calculation structure described in Sect. 4.1. Among the statistical bases, *mortality assumptions* are required for assessing the benefits (in particular in terms of the relevant expected present value) in all the multi-year health insurance covers, but in those financed with natural premiums. Further, mortality assumptions are needed for determining level premiums. Various examples will be provided in Chaps. 5 and 6.

Mortality assumptions are usually expressed in terms of a life table or a mortality law; the Heligman–Pollard law presented in Sect. 2.3.1 constitutes an example.

When disability annuities are concerned, more complex biometric models and assumptions are needed, in particular allowing for different age-patterns of mortality related to active people and disabled people respectively. Of course, disablement and recovery assumptions are also required.

As regards the choice of the life tables, or the parameters of the mortality law, it is worth stressing that health-related benefits are due in case of life. Hence, survival probabilities should not be underestimated (for both active and disabled people) in order to obtain a prudential assessment of the age-patterns of mortality.

Long-term (and, in particular, lifelong) health insurance covers should allow for future mortality trend. Hence, projected life tables or mortality laws should be adopted for pricing and reserving. Nonetheless, future mortality trend is unknown, and this implies the presence of the aggregate longevity risk (see Sect. 2.3.2). Further, more complex longevity issues affect LTCI products. As these products pay benefits to people in senescent disability condition, trends in both the total life expectancy and the disability-free life expectancy should be considered. This issue will briefly be addressed in Sect. 6.16.3.

In multi-year covers the time-value of the money should be accounted for. An *interest rate* assumption is then required to assess the benefits, as in life insurance products. As already noted, a reserving process and a consequent asset accumulation works (but in the case of natural premiums). Assets backing the reserves can provide an extra-return (with respect to the guaranteed interest rate), which can partly be credited to the policyholders. However, it is worth noting that, unlike in life insurance products with a significant saving component, in many health insurance products the reserving process implies rather modest amounts.

The reserve, backed by assets, represents a "debt" of the insurer. The debt position is a consequence of the premium arrangement. However, "abnormal" premium arrangements could lead to an insurer's credit position. This occurs, in particular, if the premiums paid by the policyholder are lower than the natural premiums (which represent the insurer's expected annual costs) in the first policy years. Premium arrangements leading to a credit position are unfeasible, and must then be avoided. Actually, a credit position would imply practical difficulties in obtaining the payment if the policyholder "lapses" (i.e. abandons) the contract. The debt position is thus required (and the requirement is denoted as the *financing condition*), and can be achieved through appropriate premium arrangements.

The effectiveness of a health insurance cover can be reduced because of inflation, e.g. because of a decreasing purchase power of a disability annuity. The quality of a multi-year cover can then be improved by adopting an indexing mechanism; this way, inflation-linked benefits can be obtained. In the case of a level premium arrangement and consequent reserving process, the benefit indexing must rely on a specific adjustment model, whose basic features are shared by life insurance and health insurance actuarial models. An example will be provided in Sect. 5.4.

4.3 Suggestions for Further Reading

Detailed references regarding actuarial features of health insurance products will be listed in Sects. 5.6 and 6.18. We now focus on some basic textbooks which can provide the reader with a general outlook on the calculation of premiums and reserves for health insurance covers.

Various aspects of health insurance (and insurances of the person in general), technical features included, are discussed in Black and Skipper (2000).

The textbooks by Bartleson (1968) and O'Grady (1988) are devoted to health insurance in general, with a special focus on sickness benefits. These benefits constitute the object of the textbook by Milbrodt (2005).

Actuarial models for disability insurance as well as several practical approaches to premium and reserve calculations are described in Haberman and Pitacco (1999), where other insurance products in the context of health insurance are also presented (i.e. several LTCI products and Critical Illness covers).

The textbook by Benjamin and Pollard (1993) deals with statistical issues in the field of life and health insurance.

The impact of various risk factors (gender in particular), either adopted as rating factors or disregarded in the pricing process, is addressed by Group Consultatif Actuariel Européen (2011) and by Riedel (2006).

We finally cite the paper by Haberman (1996) which provides an interesting insight into the evolution of actuarial science and relevant applications to life and non-life insurance problems, up to 1919; both methodological issues and practical calculation methods for disability annuities are in particular addressed. The work by Haberman and Sibbett (1995), which includes the paper just mentioned, collects the texts of numerous original contributions.

Chapter 5
Actuarial Models for Sickness Insurance

5.1 Introduction

We describe the basic actuarial structure of sickness insurance products, whose main characteristics have been presented in Sect. 3.4. In particular, we focus on insurance products which provide:

1. a fixed daily benefit in the case of (short-term) disability;
2. a hospitalization benefit, that is, a fixed daily benefit during hospital stays;
3. medical expenses reimbursement.

Products of type 1 and 2 obviously have similar technical features and, in particular, the same actuarial structure, so that we can simply refer to both of them under the label "fixed daily benefit".

We first address insurance products with a one-year cover period, then we move to products providing a multi-year cover (and possibly a lifelong cover).

5.2 One-Year Covers

As mentioned in Chap. 4, premium calculation for one-year covers has "non-life" technical features. These features are combined with "life" insurance characteristics in multi-year sickness insurance covers (as we will see in Sect. 5.3).

5.2.1 Notation and Assumptions

We adopt the following notation.

- N = random number of claims for the generic insured, within the one-year cover period, with possible outcomes $0, 1, 2, \dots$. The number N is also called the random *claim frequency*.

E. Pitacco, *Health Insurance*, EAA Series,
DOI 10.1007/978-3-319-12235-9_5

- X_j = random amount of the insured's j-th claim (e.g. medical expenses).
- Y_j = insurer's random payment for the j-th claim, called random *claim amount* or *claim severity*, given by a function of X_j, such that $Y_j \leq X_j$, reflecting the policy conditions. As regards medical expense reimbursement, an example of a function is given by Eq. (3.4.2) (referred to each claim), which accounts for flat deductible, proportional deductible and stop-loss limit.
- S = random total annual payment to the generic insured, or random *aggregate claim amount*. In general, we have:

$$S = \Phi(N; Y_1, Y_2, \ldots, Y_N). \tag{5.2.1}$$

In particular, if no policy conditions apply to the total annual amount of benefit, we have:

$$S = \begin{cases} 0 & \text{if } N = 0, \\ Y_1 + Y_2 + \cdots + Y_N & \text{if } N > 0. \end{cases} \tag{5.2.2}$$

In what follows we adopt definition (5.2.2).

For the premium calculation we assume the *equivalence principle*. Hence, the *equivalence premium* is given by the expected value of the total annual payment to the generic insured:

$$\Pi = \mathbb{E}[S] \tag{5.2.3}$$

or (to approximately take into account the timing of payments):

$$\Pi = \mathbb{E}[S]\,(1+i)^{-\frac{1}{2}}, \tag{5.2.4}$$

where i is the interest rate.

As regards $\mathbb{E}[S]$, the following assumptions are usually accepted:

1. the random variables $X_1, X_2, \ldots, X_n$ are independent of the random number N;
2. whatever the outcome n of N, the random variables $X_1, X_2, \ldots, X_n$ are
   a. mutually independent;
   b. identically distributed, and hence with a common expected value, say $\mathbb{E}[X_1]$.

We further assume that:

3. the same policy conditions are applied to all the claims, that is, $Y_j = \varphi(X_j)$ for $j = 1, 2, \ldots, n$; hence, the random variables $Y_1, Y_2, \ldots, Y_n$ are also identically distributed. It follows, in particular, that if we assign the probability distribution of Y_1, we also hold the probability distribution of any random variable Y_j; the random variables $Y_1, Y_2, \ldots, Y_n$, then, have a common expected value, say, $\mathbb{E}[Y_1]$.

Thanks to such assumptions, the expected aggregate claim amount $\mathbb{E}[S]$ can be factorized as follows:

$$\mathbb{E}[S] = \mathbb{E}[Y_1]\,\mathbb{E}[N]. \tag{5.2.5}$$

We omit the proof of this (well-known) result. However, we note that, although frequently adopted in the non-life insurance technique, the assumptions described above may be rather unrealistic. For example, the assumption of independence between the random variables X_j (and then the random variables Y_j) and the random number N may conflict with those situations in which a very high total number of claims (for instance sickness spells) is likely associated to a prevailing number of claims implying small amounts (for example because of short durations).

The quantities $\mathbb{E}[Y_1]$ (the expected claim severity) and $\mathbb{E}[N]$ (the expected claim frequency), and the interest rate i if formula (5.2.4) is adopted, constitute the *technical basis* for premium calculation.

5.2.2 From Equivalence Premiums to Gross Premiums

From Eq. (5.2.3) it clearly emerges that the random profit from the generic policy, $\Pi - S$, has an expected value equal to zero:

$$\mathbb{E}[\Pi - S] = \Pi - \mathbb{E}[S] = 0. \tag{5.2.6}$$

Thus, the equivalence principle seems to be in contrast with a reasonable profit target. Further, expenses pertaining to the policy, as well as general expenses related to the portfolio, are not accounted for, since S only refers to payment of benefits.

Actually, premiums paid by policyholders are *gross premiums* (also called *office premiums*), rather than equivalence premiums. Gross premiums are determined from equivalence premiums by adding:

1. a profit loading and contingency margins facing the risk that claims (and possibly expenses) are higher than expected;
2. an expense loading, meeting various insurer's expenses.

The items under point 1 can simply be referred to as *profit/safety loading*. Indeed, if claim (and expense) actual experience within the portfolio coincides with the related expectation, these items contribute to the portfolio profit. Conversely, in the case of experience worse than expectation, the items lower the probability (and the severity) of possible portfolio losses.

Adding the profit/safety loading to the equivalence premium yields the *net premium*; see the upper part of Fig. 5.1. For example, a fixed percentage of the equivalence premium can be added to the equivalence premium itself. Whatever the formula chosen for the loading calculation, an *explicit profit/safety loading* approach is in this case adopted. This approach relies on the use of a *natural technical basis*

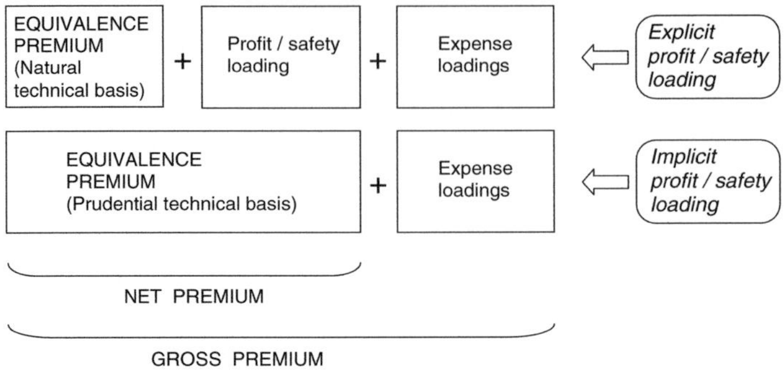

Fig. 5.1 Premium components

in the calculation of the equivalence premium, i.e. a basis which provides a realistic description of the claim (and interest) scenario. Such a basis is also referred to as the *second-order technical basis*.

The equivalence principle can also be "implemented" by adopting, instead of the natural technical basis, a *prudential technical basis* (or *safe-side technical basis*), so that the profit/safety loading is already included in the equivalence premium, then coinciding with the net premium. See the lower part of Fig. 5.1. Such a basis is also referred to as a *first-order technical basis*. Hence, this procedure implies an *implicit profit/safety loading* approach.

A prudential technical basis should consist in an expected claim frequency and expected claim severity worse than those realistically expected within the portfolio. However, it is worth stressing that the implicit loading approach is much more common in life insurance, for pricing insurance products providing benefit in the case of death, because a "worse" (and hence prudential) mortality assumption can be immediately found by referring to population life tables instead of life tables relying on portfolio (or market) experience.

"Combined" solutions are also feasible: a "weak" prudential basis can be chosen, and an explicit profit/safety loading is then added in order to determine the net premium.

With reference to item 2, i.e. the expense loading, it is worth noting that, in sickness insurance products, the expense structure is more complicated than in life insurance. Expenses are incurred not only in the underwriting phase but also in the event of a claim, and result in particular in the cost for assessing the claim itself.

As we are focussing on the basic actuarial structures, in what follows we will only refer to the calculation of equivalence premiums, hence disregarding both expense loadings and explicit profit/safety loadings.

5.2.3 Statistical Estimation

In this section we illustrate some quantities which can be used to estimate the expected claim frequency, $\mathbb{E}[N]$, the expected claim severity, $\mathbb{E}[Y_1]$, and then the expected total payout for a policy, $\mathbb{E}[S]$. Data are collected from a set of policies with specified features.

We refer to a homogeneous portfolio, consisting of r policies (i.e. insured risks), all issued at the same time and all with a one year term. Homogeneity of the policies means, in particular, that they are similar in respect of:

- the type of risk covered (e.g. either medical expense reimbursement, or fixed daily benefit);
- the policy conditions (e.g. deductibles, stop-loss limits);
- the propensity to incur into a claim;
- the possible severity of a claim, and so on.

The policy year is the same for all the policies, so that we can easily collect data on this basis. We stress that all the policies are exposed for one year (the common policy period) to the risk of incurring into one or more claims.

First, we consider a portfolio of policies providing medical expense reimbursement. Assume that, during the (policy) year, policies report z claims in total, $z \lesseqgtr r$, with claim amounts $y_1, y_2, \dots, y_z$. Note that the information is aggregate, as we just know that z claims have been reported in the portfolio, while we do not know which policies have reported such claims.

The ratio between the total payout for the portfolio and the number of policies, i.e. the *claim amount per policy*,

$$Q = \frac{y_1 + y_2 + \cdots + y_z}{r}, \tag{5.2.7}$$

is also called the *risk premium* or the *average claim cost*. Should each policy have paid a (net) premium $\Pi = Q$, then the insurer would be on balance, as the total inflow amount would be $r\,Q$, the same as the outflow amount, $y_1 + y_2 + \cdots + y_z$; for this reason, the quantity Q is regarded as an "observed premium".

The quantity Q provides an estimate of $\mathbb{E}[S]$. It is interesting to split Q as follows. The ratio

$$\bar{n} = \frac{z}{r} \tag{5.2.8}$$

represents the *average number of claims per policy*, or the *average claim frequency*. Conversely, the ratio

$$\bar{y} = \frac{y_1 + y_2 + \cdots + y_z}{z} \tag{5.2.9}$$

represents the *average claim amount per claim*, or the *average claim severity*. Note, in particular, that $\bar{n}$ expresses an estimate of $\mathbb{E}[N]$, while $\bar{y}$ provides an estimate of $\mathbb{E}[Y_1]$. Then we have

$$Q = \bar{y}\,\bar{n}, \tag{5.2.10}$$

which is the statistical version of (5.2.5).

We now focus on a portfolio of *fixed daily benefit policies*. Again, let r denote the number of insured risks and z the number of claims in the portfolio. Each claim has its own length or duration (the sickness spell, or the hospitalization period). Let $d_1, d_2, \dots, d_z$ denote the lengths of the z claims (measured in days). The quantity

$$\mu = \frac{d_1 + d_2 + \cdots + d_z}{r} \tag{5.2.11}$$

represents the average length of claim per policy, also called the *morbidity coefficient*.

We assume the same daily benefit b for all the insureds. The average claim amount per policy is then given by:

$$Q = b\,\frac{d_1 + d_2 + \cdots + d_z}{r} = b\,\mu. \tag{5.2.12}$$

The *average length per claim* is given by:

$$\bar{d} = \frac{d_1 + d_2 + \cdots + d_z}{z}. \tag{5.2.13}$$

As the average number of claims per policy is still given by Eq. (5.2.8), i.e. $\bar{n} = \frac{z}{r}$, we find

$$\mu = \bar{d}\,\bar{n}, \tag{5.2.14}$$

and hence we can split the amount Q as follows:

$$Q = b\,\mu = b\,\bar{d}\,\bar{n}. \tag{5.2.15}$$

The analogy with Eq. (5.2.10) is self-evident.

The considerations about the portfolio balance, discussed above, also hold for this type of benefit. Note that, again, $\bar{n}$ expresses an estimate of $\mathbb{E}[N]$, while $b\,\bar{d}$ provides an estimate of $\mathbb{E}[Y_1]$.

Remark 1 In a more general (and realistic) setting, the statistical procedure should take into account:

- the amounts exposed to risk (determined by allowing for policy conditions), if policies which reimburse medical expenses are concerned;

- the exposure time (within one observation year), for both medical expenses reimbursement and fixed daily benefit policies.

These aspects are beyond the scope of this textbook; the interested reader can refer to the bibliographic suggestions provided in Sect. 5.6. ■

Remark 2 More reliable estimates can be obtained by extending the observation over several years. In this case, claim amounts must be adjusted according to medical inflation. ■

5.2.4 Risk Factors and Rating Classes

Individuals belonging to a portfolio of sickness insurance policies constitute a rather heterogeneous population, in particular with regard to health-related risks. Indeed, individuals can have various ages, can be more or less healthy, can have a more or less risky occupation, etc.

Thus, we can recognize various individual *risk factors* (age, gender, current health conditions, occupation, and so on), which should be taken into account when estimating quantities as, for example, the average claim frequency or the average claim amount per claim.

Risk factors can be classified as follows.

- *Objective* risk factors are physical characteristics of the insured, in particular: age, gender, health records, occupation.
- Among the *subjective* risk factors, we recognize the personal attitude towards health, which determines the individual demand for medical treatments and, consequently, the application for insurance benefits.

Another relevant classification is the following one.

- *Observable* risk factors are those factors whose impact on claim frequency and claim severity can be assessed during the underwriting phase. Typical examples are age, gender, occupation, etc. Objective risk factors are usually observable factors.
- Other risk factors are *non-observable* factors (at least at the time of policy issue). A typical example is given by the personal attitude towards health. Further, the objective individual frailty cannot be observed (although related information can be drawn from the insured's health records).

Among the objective and observable risk factors, age has an important incidence on quantities interesting sickness insurance products. See Example 5.2.1.

Example 5.2.1 Table 5.1 shows the average claim frequency for various age groups $[x_h, x_{h+1}]$. Thus, each value $\bar{n}_x$, $x \in [x_h, x_{h+1}]$, refers to an age group. The quantity $\bar{n}$ is the overall average claim frequency, weighted with the sizes of the age groups. The data are drawn from statistical observations by the Italian Institute for Statistics (ISTAT). □

Table 5.1 Average claim frequency (*Source* ISTAT)

x	$100\,\bar{n}_x$	x	$100\,\bar{n}_x$
15–19	6.54	45–49	11.17
20–24	7.13	50–54	12.35
25–29	5.72	55–59	18.71
30–34	5.71	60–64	19.62
35–39	6.23	65–69	24.90
40–44	10.03		
$100\,\bar{n} = 10.48$			

By accounting for all the observable risk factors and the relevant possible values, we can split a population (for example, potential policyholders) into *risk classes*.

In principle, specific average claim frequency and claim severity should be estimated for each risk class, and hence a specific premium rate should be determined. However, the resulting premium rating structure could be considered too complex, or some premium rates too high. Moreover, some risk factors could not be admitted by the insurance regulation. Then, a first simplification is obtained by disregarding one or more risk factors.

When two or more risk classes are aggregated into one *rating class*, some insureds pay a premium higher than their "true" premium, i.e. the premium resulting from the risk classification, while other insureds pay a premium lower than their "true" premium. Thus, the equilibrium inside a rating class relies on a money transfer among individuals belonging to different risk classes. This transfer is usually called *solidarity* (among the insureds).

5.2.5 Premium Calculation

In what follows, we (explicitly) account for age only, as a risk factor. We assume that the quantities

$$\bar{y}_x,\ \bar{n}_x,\ \bar{d}_x$$

have been estimated for any integer age x, $x \in [x_{\min}, x_{\max}]$, i.e. within the insurable age range. We take these estimates as the technical basis for premium calculation.

Hence, for a medical expense reimbursement policy, we have:

$$\Pi_x = \bar{y}_x\,\bar{n}_x\,(1+i)^{-\frac{1}{2}} \tag{5.2.16}$$

and, for a daily benefit b:

$$\Pi_x = b\,\bar{d}_x\,\bar{n}_x\,(1+i)^{-\frac{1}{2}}. \tag{5.2.17}$$

If we consider just the average claim frequency as a function of the age, we have respectively:

$$\Pi_x = \bar{y}\,\bar{n}_x\,(1+i)^{-\frac{1}{2}} \tag{5.2.18}$$

and:

$$\Pi_x = b\,\bar{d}\,\bar{n}_x\,(1+i)^{-\frac{1}{2}}, \tag{5.2.19}$$

where $\bar{y}$ and $\bar{d}$ represent overall averages.

For practical reasons, it can be useful to factorize the quantities $\bar{n}_x$, $\bar{y}_x$ and $\bar{d}_x$, according to the logic of a *multiplicative model*. Then:

$$\bar{n}_x = \bar{n}\,t_x, \tag{5.2.20a}$$

$$\bar{y}_x = \bar{y}\,u_x, \tag{5.2.20b}$$

$$\bar{d}_x = \bar{d}\,v_x, \tag{5.2.20c}$$

where the quantities $\bar{n}$, $\bar{y}$ and $\bar{d}$ do not depend on age, whereas the *ageing coefficients* t_x, u_x, and v_x express the impact of the age as a risk factor.

Assuming that the specific age effect does not change throughout time, the claim monitoring can be restricted to the quantities $\bar{n}, \bar{y}, \bar{d}$ observed over the whole portfolio, so that more reliable estimates can be obtained.

Example 5.2.2 Consider a policy which provides a daily benefit $b = 100$. We assume $i = 0.02$ and the following statistical basis, derived from the graduation of ISTAT data:

$$\bar{n}_x = \bar{n}\,t_x = 0.1048 \times \left(0.272859 \times \mathrm{e}^{0.029841\,x}\right),$$

$$\bar{d}_x = \bar{d}\,v_x = 10.91 \times \left(0.655419 \times \mathrm{e}^{0.008796\,x}\right).$$

The average claim frequency, the average claim duration and the premium are shown in Table 5.2 for various ages. □

5.3 Multi-year Covers

In the following sections we focus on multi-year non-cancelable policies, whose conditions are stated at policy issue and cannot be changed throughout the whole policy duration.

Table 5.2 Average claim frequency, average claim duration and premium

x	$\bar{n}_x$	$\bar{d}_x$	Π_x
30	0.07000	9.30991	64.53
35	0.08126	9.72849	78.28
40	0.09434	10.16590	94.96
45	0.10952	10.62298	115.20
50	0.12714	11.10060	139.74
55	0.14760	11.59970	169.53
60	0.17135	12.12124	205.65
65	0.19892	12.66623	249.48
70	0.23093	13.23572	302.64

5.3.1 Some Preliminary Ideas

A multi-year cover can be financed, in particular, via:

1. a single premium;
2. natural premiums;
3. level premiums (throughout the whole policy duration).

Other premium arrangements can be conceived; for example:

4. "shortened" level premiums (i.e. level premiums payable throughout a period shorter than the policy duration);
5. stepwise level premiums.

Reasons for premium arrangements like 4 and 5 have been discussed in Sect. 1.2. For brevity, we now focus on arrangements 1 to 3 only.

According to the definition of "natural premium", arrangement 2 implies technical equilibrium on an annual basis, so that no policy reserve is required (but the premium reserve, or reserve for unearned premiums; see Sect. 5.3.4). Conversely, arrangements 1 and 3 guarantee the technical equilibrium only on the total policy duration as a whole. Hence, a policy reserve has to be maintained. Reserve calculation will be dealt with in Sects. 5.3.3 and 5.3.4. We now focus on the following problem: what about the policy reserve in the case the insured stops premium payment and withdraws from the contract? Possible policy conditions are as follows:

- the amount of the reserve is paid-out to the policyholder, that is, the reserve is *transferable* (e.g. to a new sickness insurance contract); in other words, the policyholder can *surrender* the contract;
- the amount of the reserve is retained by the insurer, and can be shared, according to a cross-subsidy principle, among the policies still in-force; thus, a policy *lapsation* simply occurs.

We note that, in the former case the reserve constitutes a *nonforfeiture benefit*, as the related amount will not be lost because of premature cessation of premium payment. On the contrary, in the latter case, assuming that the reserve is shared among the policies still in-force, the cross-subsidy mechanism is similar to the mutuality mechanism which works because of mortality among insureds.

It follows that, under an actuarial perspective, different probabilistic structures are needed in the two cases, for premium and reserve calculations. In the case of transferable reserve, the usual survival probability is required in the calculations. On the contrary, if the reserve is retained by the insurer, the probability that the policy is in-force is needed, as both mortality and lapses must be accounted for.

Obviously, the non-transferability of the reserve (which is a common practice in several insurance markets) has a premium-reducing effect.

In the following sections we will assume that the reserves are transferable.

5.3.2 Premiums

We focus on insurance policies providing either medical expense reimbursement or a fixed daily benefit. Let x denote the insured's age at policy issue, i.e. at time $t = 0$, and m the policy term. Of course, time is expressed in years.

The actuarial value of benefits at time $t = 0$ is given by:

$$\Pi_{x,m} = \sum_{h=0}^{m-1} {}_hp_x \, (1+i)^{-h} \, \Pi_{x+h}, \tag{5.3.1}$$

where ${}_hp_x$, according to the usual actuarial notation, denotes the probability, for a person of age x, of being alive at age $x + h$, and Π_{x+h} is given by Eqs. (5.2.16) and (5.2.17) for medical expense reimbursement and for fixed daily benefit respectively (or by Eqs. (5.2.18) and (5.2.19)).

According to the equivalence principle, the *single premium* coincides with the actuarial value, and this is expressed by Eq. (5.3.1).

Remark As regards the (possible) implicit safety loading, it is worth noting that sickness benefits are *living benefits*, that is, benefits are payable as long as the insured is alive (and sick as well). Thus, a safe-side assessment of the insurer's liabilities related to multi-year sickness insurance products requires that the mortality of the insureds should not be overestimated. ■

The quantities $\Pi_x, \Pi_{x+1}, \ldots, \Pi_{x+m-1}$ constitute the *natural premiums* of the m-year insurance cover. We usually find:

$$\Pi_x < \Pi_{x+1} < \cdots < \Pi_{x+m-1} \tag{5.3.2}$$

because of the age effect (see, for example, Table 5.2).

It is interesting to find the expression for the single premium in a multiplicative model (as expressed by Eqs. (5.2.20a)–(5.2.20c)). For example, if

$$\Pi_x = \bar{y}_x\, \bar{n}_x\, (1+i)^{-\frac{1}{2}} = \bar{y}\, \bar{n}\, u_x\, t_x\, (1+i)^{-\frac{1}{2}} \tag{5.3.3}$$

then we have:

$$\begin{aligned} \Pi_{x,m} &= \sum_{h=0}^{m-1} {}_h p_x\, (1+i)^{-h}\, \bar{y}_{x+h}\, \bar{n}_{x+h}\, (1+i)^{-\frac{1}{2}} \\ &= \bar{y}\, \bar{n} \sum_{h=0}^{m-1} {}_h p_x\, (1+i)^{-h-\frac{1}{2}}\, u_{x+h}\, t_{x+h} \end{aligned} \tag{5.3.4}$$

and briefly, with obvious meaning of the symbols:

$$\begin{aligned} \Pi_{x,m} &= K \sum_{h=0}^{m-1} w_{x,h} \\ &= K\, \pi_{x,m}, \end{aligned} \tag{5.3.5}$$

where K is independent of age, whereas the $w_{x,h}$'s depend on both the initial age x and the time h elapsed from policy issue.

Various premium arrangements based on a sequence of periodic premiums can be conceived. In particular, an example is given by the sequence of natural premiums $\Pi_x, \Pi_{x+1}, \ldots, \Pi_{x+m-1}$. This arrangement implies an increasing annual cost to the policyholder (see inequalities (5.3.2)).

To avoid increasing costs, *annual level premiums* can be charged. Assuming that annual level premiums, $P_{x,m}$, are payable for m years, we have:

$$P_{x,m} = \frac{\Pi_{x,m}}{\ddot{a}_{x:\overline{m}|}}, \tag{5.3.6}$$

where $\ddot{a}_{x:\overline{m}|}$, according to the usual notation, is the actuarial value of a unitary temporary life annuity payable at the beginning of each year. We find:

$$P_{x,m} = \frac{\sum_{h=0}^{m-1} {}_h p_x\, (1+i)^{-h}\, \Pi_{x+h}}{\sum_{h=0}^{m-1} {}_h p_x\, (1+i)^{-h}}. \tag{5.3.7}$$

Thus, the annual level premium (assumed payable throughout the whole policy duration) can be expressed as the arithmetic weighted average of the natural premiums.

Example 5.3.1 We consider insurance policies providing fixed daily benefits. Assume:

$i = 0.02$;
$b = 100$;
$\bar{n}_y, \bar{d}_y$ as in Example 5.2.2;

Table 5.3 Single premiums

x	$m = 5$	$m = 10$	$m = 15$	$m = 20$
30	334.86	701.78	1 103.13	1 540.82
35	406.02	850.13	1 334.46	1 859.98
40	492.11	1 028.79	1 611.12	2 237.62
45	596.11	1 242.92	1 938.80	2 676.86
50	721.35	1 497.42	2 320.53	3 172.86
55	871.42	1 795.66	2 752.71	–
60	1 049.76	2 136.79	–	–
65	1 258.68	–	–	–

Table 5.4 Annual level premiums

x	$m = 5$	$m = 10$	$m = 15$	$m = 20$
30	69.71	76.75	84.49	92.97
35	84.56	93.10	102.46	112.69
40	102.58	112.92	124.23	136.51
45	124.43	136.94	150.55	165.22
50	150.93	166.03	182.34	199.65
55	183.06	201.23	220.60	–
60	222.01	243.75	–	–
65	269.20	–	–	–

the mortality assumption is expressed by the (first) Heligman–Pollard law (with $\omega = 110$); see Sect. 2.3, and Eq. (2.3.1) in particular; parameters are as specified in Example 2.3.1.

Table 5.3 shows the single premiums for various ages x at policy issue, and policy terms m, with the constraint $x + m \leq 70$. Conversely, Table 5.4 shows the annual level premiums for the same ages at policy issue and the same policy terms.

Natural premiums are plotted in Figs. 5.2 and 5.3. In particular, natural premiums are compared to the level premium amount in Fig. 5.2, whereas the increase in natural premiums related to various ages at policy issue is shown in Fig. 5.3. □

5.3.3 Reserves

The prospective mathematical reserve, V_t, for a sickness insurance policy at (integer) time t is defined as the actuarial value of future benefits less the actuarial value of future premiums. Denoting the two actuarial values by $\text{Ben}(t, m)$ and $\text{Prem}(t, m)$, respectively, in formal terms we have:

$$V_t = \text{Ben}(t, m) - \text{Prem}(t, m); \quad t = 0, 1, \ldots, m. \tag{5.3.8}$$

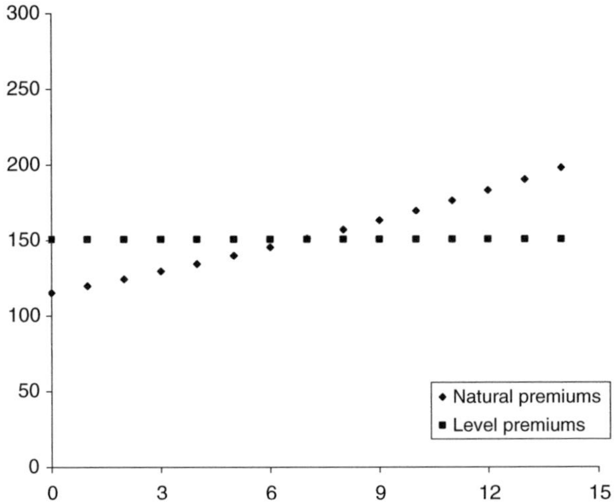

Fig. 5.2 Natural premiums and annual level premium; $x = 45, m = 15$

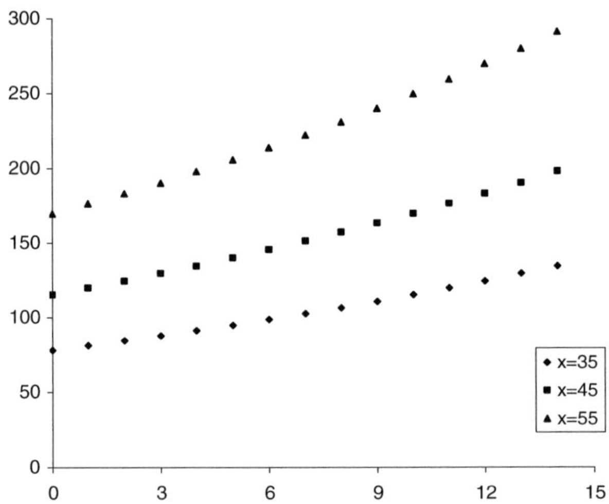

Fig. 5.3 Natural premiums for various ages at policy issue; $m = 15$

This reserve is also referred to as the *aging reserve*, or *senescence reserve*, to stress the age effect on the behavior of natural premiums and the consequent need for setting up a reserve.

Consider the case of annual level premiums payable for the whole policy duration; for simplicity, we now denote by P the annual level premium $P_{x,m}$ given by Eqs. (5.3.6) and (5.3.7). Then, Eq. (5.3.8) yields:

$$V_t = \Pi_{x+t,m-t} - P\,\ddot{a}_{x+t:\overline{m-t}|}; \quad t = 0, 1, \ldots, m. \tag{5.3.9}$$

Of course, we find:

$$V_0 = V_m = 0. \tag{5.3.10}$$

From Eq. (5.3.9), we have:

$$V_t = \Pi_{x+t,1} - P + {}_1p_{x+t}\,(1+i)^{-1}\,(\Pi_{x+t+1,m-t-1} - P\,\ddot{a}_{x+t+1:\overline{m-t-1}|}) \tag{5.3.11}$$

and, as $\Pi_{x+t,1} = \Pi_{x+t}$, we obtain the recursion:

$$V_t + P = \Pi_{x+t} + {}_1p_{x+t}\,(1+i)^{-1}\,V_{t+1}, \tag{5.3.12}$$

which expresses the technical balance in year $(t, t+1)$. In particular, V_t can be interpreted as the amount of *assets* coming from the accumulation of part of the premiums cashed before time t, whereas the quantity V_{t+1} represents the *debt* of the insurer for future benefits, net of the credit for future premiums.

Example 5.3.2 We consider insurance policies providing fixed daily benefits. The data are as in Example 5.3.1. The time profiles of the policy reserve for various ages at entry and policy terms are shown in Figs. 5.4 and 5.5 respectively. □

5.3.4 Reserves at Fractional Durations

The analysis of the time profile of the reserve has so far been restricted to the policy anniversaries, namely integer past durations (since the policy issue). The extension

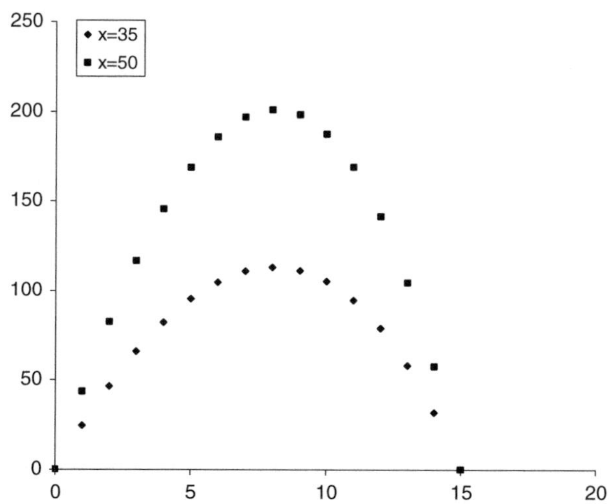

Fig. 5.4 The policy reserve for two ages at policy issue; $m = 15$

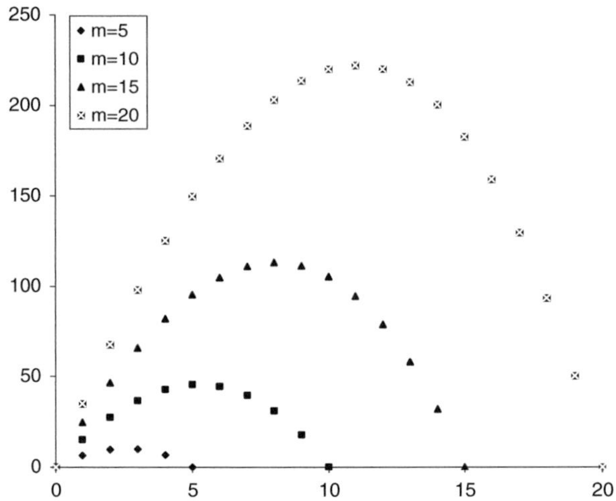

Fig. 5.5 The policy reserve for various policy terms; $x = 35$

to fractional durations is, however, of practical interest. For example, the need for calculating the policy reserve (and the portfolio reserve, as well) at times other than the policy anniversaries arises when assessing the items of technical and financial reports, and the balance-sheet in particular.

The calculation of the exact value of the policy reserve at all durations can be carried out in a time-continuous setting. In the actuarial practice, however, it is rather common to work in a time-discrete framework (as we are actually doing), and to obtain approximations to the exact value of the reserve via interpolation procedures, in particular by adopting linear interpolation formulae. Here we illustrate the interpolation approach, focussing on some examples.

Consider a sickness insurance policy with premium arrangement based on natural premiums. Thus, premiums $\Pi_x, \Pi_{x+1}, \dots, \Pi_{x+m-1}$ are cashed by the insurer at times $0, 1, \dots, m-1$ respectively. The reserve is, of course, equal to zero at all the policy anniversaries, before cashing the premium which falls due at that time; thus, $V_t = 0$ for $t = 0, 1, \dots, m-1$ (as well as $V_m = 0$). Immediately after cashing the premium, the insurer's debt (and the corresponding asset) is clearly equal to the premium itself; hence, denoting by V_{t^+} the reserve after cashing the premium, we have:

$$V_{t^+} = \Pi_{x+t}; \quad t = 0, 1, \dots, m-1. \tag{5.3.13}$$

Then, the premium is used throughout the year to contribute to the payment of benefits to (same-age) policyholders, according to the mutuality mechanism, and, again, we have $V_{t+1} = 0$. At time $t + r$, with $t = 0, 1, \dots, m-1$ and $0 < r < 1$, we let:

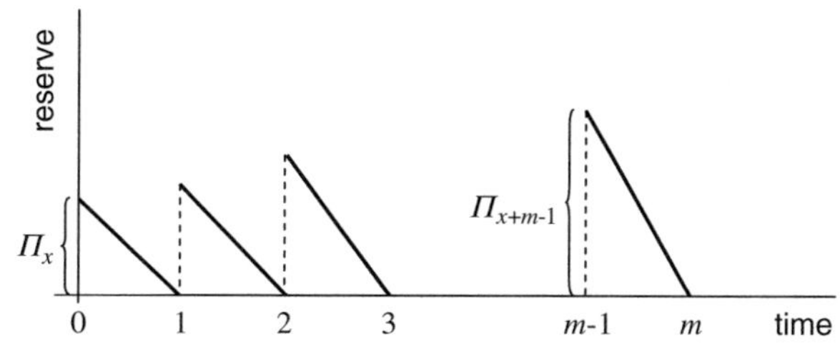

Fig. 5.6 Interpolated reserve profile in the case of natural premiums

$$V_{t+r} = (1-r)\, V_{t^+} = (1-r)\, \Pi_{x+t}. \tag{5.3.14}$$

The resulting time profile of the reserve is plotted in Fig. 5.6. We note that the reserve $V_{t+r} = (1-r)\,\Pi_{x+t}$ is an *unearned premium reserve*, according to the non-life insurance terminology.

As a second example, we refer to an insurance product with annual level premiums P. After cashing the premium, which falls due at time t, the reserve raises from V_t to

$$V_{t^+} = V_t + P; \quad t = 0, 1, \dots, m-1. \tag{5.3.15}$$

Then, the linear interpolation yields:

$$V_{t+r} = (1-r)\, V_{t^+} + r\, V_{t+1} = [(1-r)\, V_t + r\, V_{t+1}] + (1-r)\, P. \tag{5.3.16}$$

See Fig. 5.7. The term $(1-r)\,P$ represents the unearned premium reserve. We note, in particular, the following aspects.

- Interpolating between V_t (instead of V_{t^+}) and V_{t+1} would cause an apparent underestimation of the reserve at all times between t and $t+1$ (again, see Fig. 5.7).

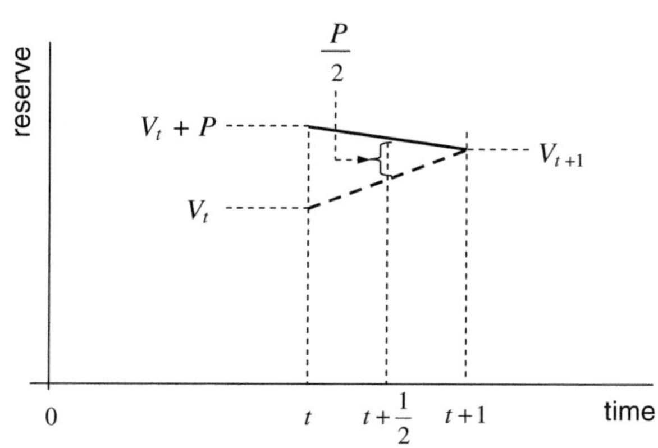

Fig. 5.7 Reserve interpolation in the case of annual level premiums

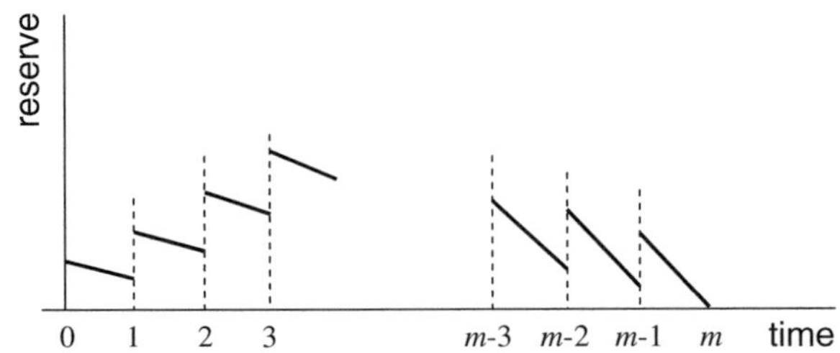

Fig. 5.8 Interpolated reserve profile in the case of annual level premiums

- The "use" of the premium P changes throughout time. In particular, the share of the premium used to cover sickness benefits according to the mutuality mechanism is increasing throughout the policy duration, because the expected number of claims (and possibly the expected claim amount) increases as the policyholders' age increases. This fact determines a time profile of the reserve like the one plotted in Fig. 5.8, in which we can see the increasing (negative) slope of the interpolating segments.

As a third example, we consider an insurance product with a single premium $\Pi_{x,m}$. In this case, there is no jump in the reserve profile, but at the payment of the single premium, when the reserve jumps from $V_0 = 0$ to $V_{0^+} = \Pi_{x,m}$. Then, the linear interpolation procedure is as follows:

$$V_r = (1 - r)\, V_{0^+} + r\, V_1, \tag{5.3.17a}$$

$$V_{t+r} = (1 - r)\, V_t + r\, V_{t+1} \quad \text{for } t = 1, 2, \ldots, m - 1. \tag{5.3.17b}$$

See Fig. 5.9.

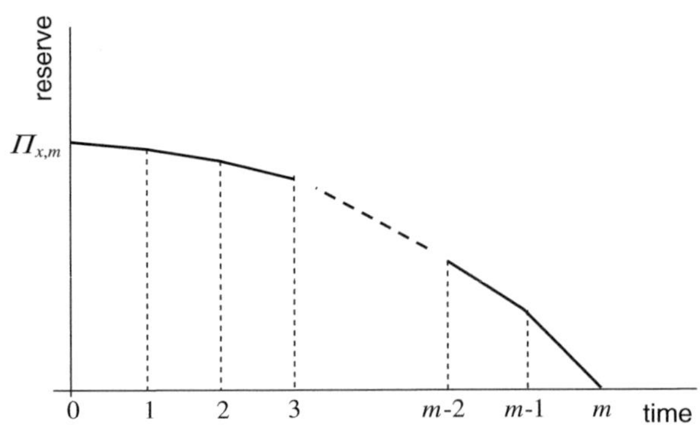

Fig. 5.9 Interpolated reserve profile in the case of a single premium

5.4 Indexation Mechanisms

In this section we refer to multi-year covers (and, in particular, to long-term covers), which are exposed to the risk that significant changes in some scenario elements jeopardize the technical balance between benefits and premiums. Some approaches to this risk are sketched, and one specific approach is described and discussed in Sects. 5.4.2 and 5.4.3.

5.4.1 Introduction

We have so far assumed a "static" scenario, in which no significant change occurs, throughout the policy duration, in the elements accounted for at the time of policy issue, for example:

- the expected claim frequency
- the expected claim severity
- the money purchase power
- the mortality assumption, etc.

If changes do occur, e.g. in the expected claim frequency or the expected claim severity, the technical equilibrium between premiums and benefits can be jeopardized, or, if a fixed daily benefit is concerned and its amount is kept constant, the effectiveness of the insurance cover can be reduced because of inflation.

To avoid, or to limit to some extent, the consequences of non-equilibrium situations, two basic approaches can be adopted.

1. A forecast of future trend in some elements (for example, the expected claim severity, or the inflation rate) can be introduced into the benefit assessment and hence into the premium and reserve calculations; of course, no guarantee can be provided as regards the effectiveness of such an approach.
2. A periodic (e.g. yearly) *a posteriori* adjustment procedure can be adopted, which first consists in re-assessing the benefits according to the observed scenario, and then:

    a. either re-determining the future premiums and/or the reserve needed to maintain the technical balance according to the equivalence principle,
    b. or changing some policy conditions; for example, with reference to medical expense reimbursement, raising the proportional deductible, still to maintain the technical balance.

Of course, "combined" solutions, based on approaches 1, 2a and 2b, can be conceived and implemented. In what follows, we only focus on solutions of type 2a.

5.4.2 The Adjustment Model

We refer to an insurance policy with annual level premiums payable throughout the whole policy duration. We stress that the expression “level premiums” always refers to the initial calculation of premiums, whereas the actual annual premiums will be progressively redetermined according to the adjustment mechanism.

We assume that at time t ($t = 1, 2, \dots, m-1$) an adjustment in the future premiums and/or the reserve can occur due to a reassessment of the value of future benefits. We now use the following notation to denote quantities referred at time t before the (possible) adjustment at that time, but including the (possible) adjustments up to time $t-1$:

- V_{t^-} = the policy reserve;
- $\mathrm{Ben}(t^-, m)$ = the actuarial value of future benefits;
- $\mathrm{Prem}(t^-, m)$ = the actuarial value of future premiums.

The technical balance is then expressed by the following relation:

$$V_{t^-} + \mathrm{Prem}(t^-, m) = \mathrm{Ben}(t^-, m) \tag{5.4.1}$$

(see Eq. (5.3.8)).

Assume now that at time t the actuarial value of future benefits is adjusted, so that it increases at the rate $j^{[\mathrm{B}]}(t)$. To maintain the actuarial balance, the total quantity in the left-hand side of (5.4.1) must also be increased at the rate $j^{[\mathrm{B}]}(t)$. Thus, the new balance condition is expressed as follows:

$$\left(V_{t^-} + \mathrm{Prem}(t^-, m)\right)\left(1 + j^{[\mathrm{B}]}(t)\right) = \mathrm{Ben}(t^-, m)\left(1 + j^{[\mathrm{B}]}(t)\right). \tag{5.4.2}$$

Condition (5.4.2) does not require that both the reserve and the future premiums are increased at the rate $j^{[\mathrm{B}]}(t)$; what is required is that their total value is increased at the rate $j^{[\mathrm{B}]}(t)$. Different rates of adjustment of the reserve and the future premiums, respectively denoted by $j^{[\mathrm{V}]}(t)$ and $j^{[\mathrm{P}]}(t)$, can be adopted, provided that the following balance condition is satisfied:

$$V_{t^-}\left(1 + j^{[\mathrm{V}]}(t)\right) + \mathrm{Prem}(t^-, m)\left(1 + j^{[\mathrm{P}]}(t)\right) = \mathrm{Ben}(t^-, m)\left(1 + j^{[\mathrm{B}]}(t)\right). \tag{5.4.3}$$

Since (5.4.1) must be fulfilled, Eq. (5.4.3) requires:

$$V_{t^-}\, j^{[\mathrm{V}]}(t) + \mathrm{Prem}(t^-, m)\, j^{[\mathrm{P}]}(t) = \mathrm{Ben}(t^-, m)\, j^{[\mathrm{B}]}(t) \tag{5.4.4}$$

(as we obtain by subtracting (5.4.1) from (5.4.3)). Equation (5.4.4) expresses that the adjustments of the reserve and premiums must be on actuarial balance with the benefit adjustment.

Equation (5.4.4) admits an infinite number of solutions: indeed, given the value of $j^{[\mathrm{B}]}(t)$, the two unknowns are $j^{[\mathrm{V}]}(t)$ and $j^{[\mathrm{P}]}(t)$. The effectiveness of specific solutions will be discussed in Sect. 5.4.3 and, in particular, in Example 5.4.1.

It is interesting to obtain an expression for $j^{[B]}(t)$ from Eq. (5.4.4). We find:

$$j^{[B]}(t) = \frac{V_{t^-}\, j^{[V]}(t) + \mathrm{Prem}(t^-, m)\, j^{[P]}(t)}{\mathrm{Ben}(t^-, m)} \tag{5.4.5}$$

and, replacing $\mathrm{Ben}(t^-, m)$ according to (5.4.1), we obtain:

$$j^{[B]}(t) = \frac{V_{t^-}\, j^{[V]}(t) + \mathrm{Prem}(t^-, m)\, j^{[P]}(t)}{V_{t^-} + \mathrm{Prem}(t^-, m)}. \tag{5.4.6}$$

From Eq. (5.4.6) we see that the rate of adjustment of the benefits, $j^{[B]}(t)$, is a weighted arithmetic mean of the rate of adjustment of the reserve, $j^{[V]}(t)$, and the rate of adjustment of premiums, $j^{[P]}(t)$. The weights, respectively $\dfrac{V_{t^-}}{V_{t^-} + \mathrm{Prem}(t^-, m)}$ and $\dfrac{\mathrm{Prem}(t^-, m)}{V_{t^-} + \mathrm{Prem}(t^-, m)}$, change in time. At any time,

$$\begin{aligned} &\text{if } j^{[V]}(t) < j^{[B]}(t), \text{ then } j^{[P]}(t) > j^{[B]}(t);\\ &\text{if } j^{[P]}(t) < j^{[B]}(t), \text{ then } j^{[V]}(t) > j^{[B]}(t). \end{aligned}$$

The reserve to be set up at time t, after the benefit adjustment (but before the premium payment), is:

$$V_t = V_{t^-} \left(1 + j^{[V]}(t)\right). \tag{5.4.7}$$

From Eq. (5.4.3), we obtain the relevant expression in terms of actuarial value of future benefits and premiums, namely

$$V_t = \mathrm{Ben}(t^-, m) \left(1 + j^{[B]}(t)\right) - \mathrm{Prem}(t^-, m) \left(1 + j^{[P]}(t)\right). \tag{5.4.8}$$

If we let

$$\mathrm{Ben}(t, m) = \mathrm{Ben}(t^-, m) \left(1 + j^{[B]}(t)\right), \tag{5.4.9a}$$

$$\mathrm{Prem}(t, m) = \mathrm{Prem}(t^-, m) \left(1 + j^{[P]}(t)\right), \tag{5.4.9b}$$

we can also write:

$$V_t = \mathrm{Ben}(t, m) - \mathrm{Prem}(t, m), \tag{5.4.10}$$

which shows us that V_t is a prospective reserve, as is desirable, given that the adjustment has not been motivated by the need of revising the logic for the calculation of the reserve. Note that the reserve V_t is assessed taking into account the updated benefit and premium amounts, while the technical basis is unchanged.

5.4.3 Application to Sickness Insurance Covers

In this section, we show how the adjustment model described in Sect. 5.4.2 applies to sickness insurance covers.

We refer to a sickness insurance policy, issued at time 0, with maturity m, and annual level premium $P_{x,m}$ (calculated at policy issue). The benefit can either consist in medical expense reimbursement or a fixed daily benefit. In both cases, the annual level premium is then given by Eq. (5.3.6), with $\Pi_{x,m}$ given in general by (5.3.1).

The policy is designed so that at each policy anniversary it is possible to adjust the benefits, following (5.4.6). We then need to extend the notation, in line with the one adopted in Sect. 5.4.2. First, we note that, obviously,

$$\mathrm{Ben}(0, m) = \Pi_{x,m}, \tag{5.4.11a}$$

$$\mathrm{Prem}(0, m) = P_{x,m}\, \ddot{a}_{x:\overline{m}|}. \tag{5.4.11b}$$

As regards the premiums, let

$$P(0) = P_{x,m} \tag{5.4.12}$$

and then:

$$P(t) = P(t-1)\left(1 + j^{[\mathrm{P}]}(t)\right). \tag{5.4.13}$$

We assume that the actuarial value of the benefits can be expressed by the multiplicative model (see Eq. (5.3.5)). We further assume that the possible changes only affect the factor K (which is independent of age), and do not concern the specific effect of age. Let $K(0)$ denote the value of the factor at policy issue. Hence (see Eqs. (5.3.4) and (5.3.5)),

$$\Pi_{x,m} = K(0)\, \pi_{x,m} \tag{5.4.14}$$

and then:

$$K(t) = K(t-1)\left(1 + j^{[\mathrm{B}]}(t)\right). \tag{5.4.15}$$

For a policy providing medical expense reimbursement, we have:

$$K(0) = \bar{y}(0)\, \bar{n}, \tag{5.4.16}$$

where $\bar{y}(0)$ denotes the expected claim severity assessed at the policy issue. If the inflation affects the expense amount (whereas the expected claim frequency is assumed to be constant over time), we have:

$$K(t) = \bar{y}(t)\, \bar{n} = \bar{y}(t-1)\left(1 + j^{[\mathrm{B}]}(t)\right) \bar{n}, \tag{5.4.17}$$

with the obvious meaning of $\bar{y}(t)$ and $\bar{y}(t-1)$.

For a policy providing a daily benefit, we have:

$$K(0) = b(0)\,\bar{d}\,\bar{n}, \tag{5.4.18}$$

where $b(0)$ denotes the initial value of the daily benefit. Its amount can be changed to keep the purchasing power; then:

$$K(t) = b(t)\,\bar{d}\,\bar{n} = b(t-1)\,\big(1 + j^{[B]}(t)\big)\,\bar{d}\,\bar{n}. \tag{5.4.19}$$

We note that:

$$\mathrm{Ben}(t^-, m) = K(t-1)\,\pi_{x+t,m-t}, \tag{5.4.20a}$$

$$\mathrm{Prem}(t^-, m) = P(t-1)\,\ddot{a}_{x+t:\overline{m-t}|}, \tag{5.4.20b}$$

so that Eqs. (5.4.9a) and (5.4.9b) can finally be applied to find the reserve V_t.

It is worth noting that, in practice, the increase in the reserve (at the rate $j^{[V]}(t)$) is usually financed by the insurer via profit participation, whereas the increase in premiums (at the rate $j^{[P]}(t)$) is paid by the policyholder.

Example 5.4.1 We refer to a policy providing medical expense reimbursement. We assume $x = 50$, $m = 15$. Annual level premiums are payable for the whole policy duration. As regards other data, see Examples 5.2.2 and 5.3.1. Table 5.5 shows the annual benefit increments which can be maintained by a 5 % annual increment of the reserve, while the premium remain unchanged. Benefit increments are generally low, and extremely low in the first policy years, i.e. when the reserve is very small and many premiums have still to be paid; formally, this fact can be easily explained

Table 5.5 Benefit adjustments (1)

t	$j^{[B]}(t)$	$j^{[V]}(t)$	$j^{[P]}(t)$
1	0.00098	0.05	0
2	0.00198	0.05	0
3	0.00301	0.05	0
4	0.00407	0.05	0
5	0.00515	0.05	0
6	0.00625	0.05	0
7	0.00736	0.05	0
8	0.00850	0.05	0
9	0.00965	0.05	0
10	0.01081	0.05	0
11	0.01198	0.05	0
12	0.01316	0.05	0
13	0.01434	0.05	0
14	0.01552	0.05	0

Table 5.6 Benefit adjustments (2)

t	$j^{[B]}(t)$	$j^{[V]}(t)$	$j^{[P]}(t)$
1	0.06	0	0.06120
2	0.06	0	0.06234
3	0.06	0	0.06345
4	0.06	0	0.06450
5	0.06	0	0.06550
6	0.06	0	0.06646
7	0.06	0	0.06737
8	0.06	0	0.06823
9	0.06	0	0.06905
10	0.06	0	0.06982
11	0.06	0	0.07055
12	0.06	0	0.07123
13	0.06	0	0.07187
14	0.06	0	0.07247

by looking at Eq. (5.4.6). Conversely, Table 5.6 shows the annual increments in the premium required to maintain a 6 % annual increment in the benefit value (without any increment in the reserve, that is, without any contribution from the insurance company).

It is worth noting that sickness insurance covers are not "accumulation" products; hence, the amount of the mathematical reserve is relatively small (although the longer the policy duration, the higher the mathematical reserve; see the numerical results in Example 5.3.2). Then:

- the only increment of the reserve cannot maintain a sensible raise in the actuarial value of future benefits (see Table 5.5);
- on the contrary, the raise in the actuarial value of future benefits can be financed by a reasonable increment of future premiums only (see Table 5.6);
- a longer policy term implies higher reserve amounts (see, for example, Fig. 5.5), and hence a more important role of the reserve increments in maintaining the raise in the actuarial value of future benefits. □

5.5 Lifelong Covers

The actuarial value of the benefits provided by a lifelong sickness cover is given by:

$$\Pi_{x,\infty} = \sum_{h=0}^{+\infty} {}_h p_x \, (1+i)^{-h} \, \Pi_{x+h}, \tag{5.5.1}$$

where the Π_{x+h}'s are given by Eqs. (5.2.16) (or (5.2.18)) and (5.2.17) (or (5.2.19)), according to the type of benefit.

Several periodic premium arrangements can be conceived and implemented. For comments regarding the feasibility of various premium arrangements, the reader can refer to Sect. 1.2.4. The premium arrangement exclusively based on natural premiums (see Fig. 1.4) does not imply any actuarial problem; conversely, it is worth noting that at very old ages the natural premiums can be extremely high, and this represents an obstacle to the applicability of this arrangement. Hence, we focus on the following arrangements:

1. lifelong level premiums;
2. temporary level premiums;
3. temporary stepwise level premiums.

If arrangement 1 is adopted (see Fig. 1.6), the annual level premium, $P_{x,\infty(\infty)}$, is given by:

$$P_{x,\infty(\infty)} = \frac{\Pi_{x,\infty}}{\ddot{a}_x}, \tag{5.5.2}$$

whereas arrangement 2 (see Fig. 1.7), assuming that r premiums are payable, leads to:

$$P_{x,\infty(r)} = \frac{\Pi_{x,\infty}}{\ddot{a}_{x:\overline{r}|}}. \tag{5.5.3}$$

As regards arrangement 3 (see Fig. 1.8), we again assume that r premiums are payable, and, in particular, premium P' is payable for r' years, premium P'' for the following r'' years, and premium P''' for the remaining r''' years (of course $r' + r'' + r''' = r$). Premiums must fulfill the equivalence principle expressed by the following equation:

$$P' \, \ddot{a}_{x:\overline{r'}|} + P'' \, {}_{r'|}\ddot{a}_{x:\overline{r''}|} + P''' \, {}_{r'+r''|}\ddot{a}_{x:\overline{r'''}|} = \Pi_{x,\infty}. \tag{5.5.4}$$

To solve Eq. (5.5.4), a relation among P', P'', P''' must be assigned (reasonably, such that $P' < P'' < P'''$). Further, the financing condition must be fulfilled (see Sect. 4.2.2).

Some remarks follow, concerning the features of a lifelong sickness cover.

- From the individual perspective, a lifelong sickness insurance policy provides the insured with appropriate coverage over his/her whole life, and hence it can be regarded as a "high quality" insurance product.
- From the insurers' perspective, some problems may arise; in particular:
  - sickness data related to very old ages may be scanty, in particular in those insurance markets in which lifelong sickness covers constitute a new issue;
  - the need for forecasting mortality (and morbidity) over very long periods implies significant aggregate longevity risk (see Sect. 2.3);

- as seen in Sect. 5.3.3, the longer the policy duration the higher is the reserve (see in particular Fig. 5.5); then, the investment of assets backing the reserve gains in importance;
- at the same time, higher reserve amounts imply a more important role of the reserve itself in maintaining indexing mechanisms (see the comments in Example 5.4.1).

5.6 Suggestions for Further Reading

Papers, books and reports dealing with general aspects of sickness insurance have been considered in Sect. 3.13. The reader can refer to the relevant citations as regards descriptions of insurance products and related policy conditions, financing care systems, comparisons among local health care systems in various countries, etc. In this section we only focus on papers, reports and textbooks addressing actuarial aspects of sickness insurance.

The following textbooks deal with both general and actuarial aspects of sickness insurance: Bartleson (1968), Black and Skipper (2000), Milbrodt (2005), and O'Grady (1988).

Transferable aging reserves in sickness insurance policies are addressed by Baumann et al. (2004). A survey of private sickness insurance in Germany, including actuarial aspects related to premium and reserve calculations, is provided by Fürhaupter and Brechtmann (2002).

The paper by Olivieri and Pitacco (2002b) focusses on the impact of the (aggregate) longevity risk on the management of portfolios of lifelong sickness covers; various pricing solutions are suggested and compared. The actuarial values of several health insurance products, including sickness covers, and related approximation formulae are analyzed in Olivieri and Pitacco (2009).

Medical expense insurance and related actuarial aspects are specifically addressed by Orros and Webber (1988). Provisioning, pricing and forecasting in sickness insurance are focussed by Lurie (2007). Rating systems allowing for *a posteriori* risk classification, relying on individual claim experience, are analyzed in Pitacco (1992). The paper by Vercruysse et al. (2013) deals with indexation mechanisms in lifelong sickness insurance products; various solutions, which aim at sharing the cost of indexation between insurer and policyholders, are analyzed.

Chapter 6
Actuarial Models for Disability Annuities

6.1 Introduction

The main features of disability insurance products have been described in Sect. 3.5. In this chapter we specifically refer to insurance products providing disability annuities, starting with a (very) basic type of annuity, described in Sect. 6.2.

In the following sections, we focus on a simple probabilistic model (Sect. 6.3) and the calculation of actuarial values (Sect. 6.4). We then address premium and reserve calculation (Sects. 6.5 and 6.6); some numerical examples are also provided.

More realistic products are described, in terms of the relevant policy conditions, in Sect. 6.7, while more complex modelling issues are dealt with in Sects. 6.8 and 6.9.

Sections 6.10–6.15 focus on calculation methods adopted in the actuarial practice, while Sect. 6.16 provides an introduction to actuarial models for LTCI products, by extending the model adopted for disability annuities.

Finally, some basic ideas concerning the representation of the mortality among disabled people are presented in Sect. 6.17.

6.2 Some Preliminary Ideas

Let us consider an insurance cover providing a disability annuity benefit b per annum when the insured is disabled, i.e. in state i. At policy issue the insured, aged x, is active, i.e. in state a. The policy term is m. The disability annuity is assumed to be payable up to the end of the policy term m. For simplicity, we assume that the benefit is paid at policy anniversaries (see Fig. 6.1). This assumption is rather unrealistic, but the resulting model is simple and allows us to single out important basic ideas. No particular policy condition (e.g. deferred period, waiting period, etc.) is considered. More realistic assumptions lead to much more complicated models (see, in particular, Sects. 6.7 and 6.11).

E. Pitacco, *Health Insurance*, EAA Series,
DOI 10.1007/978-3-319-12235-9_6

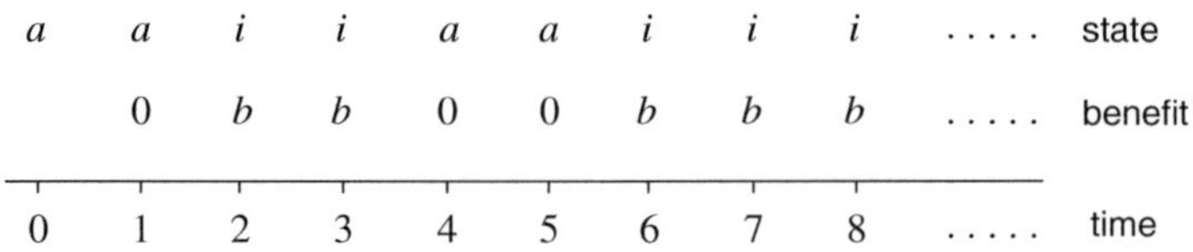

Fig. 6.1 An example of disability annuity

The annual amount paid by the insurer to the insured at time h is then a random amount, B_h, defined as follows:

$$B_h = \begin{cases} b & \text{if, at time } h, \text{ state } = i, \\ 0 & \text{if, at time } h, \text{ state } \neq i. \end{cases} \tag{6.2.1}$$

The present value Y of the annuity whose amounts are B_h, for $h = 1, 2, \ldots, m$, is, in its turn, a random amount. Let v denote the annual discount factor; we then have:

$$Y = \sum_{h=1}^{m} B_h \, v^h. \tag{6.2.2}$$

Hence, the expected present value (shortly, the *actuarial value*) is given by:

$$\mathbb{E}[Y] = \sum_{h=1}^{m} \mathbb{E}[B_h] \, v^h = \sum_{h=1}^{m} b \, {}_h p_x^{ai} \, v^h, \tag{6.2.3}$$

where ${}_h p_x^{ai}$ denotes, according to the traditional actuarial notation, the probability that the insured, aged x and active (state a) at policy issue, is disabled (state i) at age $x + h$; hence, ${}_h p_x^{ai}$ also denotes the probability that the amount b is paid at time h.

The direct implementation of formula (6.2.3) is not a trivial matter, in spite of its formal simplicity. Actually, there are practical difficulties in "directly" estimating the probabilities ${}_h p_x^{ai}$ from statistical data. So, an alternative approach to the calculation of the actuarial value is needed, in particular looking for a probabilistic structure compatible with data which are reasonably available.

6.3 The Basic Biometric Model

The evolution of an insured risk throughout time can be viewed as a sequence of events which determine the cash flows of premiums and benefits. The logical support to describe the evolution of a risk is provided by a multistate model.

6.3.1 Multistate Models for Disability Insurance

When disability insurance products are referred to, relevant events are typically disablement, recovery and death. The evolution of a risk (an insured individual) can then be described in terms of the presence of the risk itself, at every point of time, in a certain *state*, belonging to a given set of states, or *state space*. The aforementioned events correspond to *transitions* from one state to another state.

A graphical representation of a *multistate model* consisting of states and transitions is provided by a *directed graph* (or *digraph*), whose *vertices* (or *nodes*) represent the states, whereas the *edges* (or *arcs*) represent possible (direct) transitions between states. The graphs of Fig. 6.2a–c refer to disability insurance covers. The meaning of the states is as follows:

a = active (or healthy)
i = disabled (or invalid)
d = dead

The following transitions are represented:

$a \to i$ = disablement
$a \to d$ = death of an active individual
$i \to d$ = death of a disabled individual
$i \to a$ = recovery

According to specific problems and models adopted, some transitions can be excluded, as we will see in what follows.

A graph simply describes the contingencies pertaining to an insured risk, as far as its evolution throughout time is concerned. A probabilistic structure must be introduced in order to express a numerical assessment of the contingencies. A further step in describing the features of an insurance cover consists in relating premiums and benefits to the presence of the insured risk in some states or to transitions of the risk itself from one state to another.

An insurance cover just providing a lump sum benefit in the case of permanent (and total) disability can be represented by the three-state model depicted in Fig. 6.2a (also called a double-decrement model). Note that, since only permanent (and total) disability is involved, the label "active" concerns any insured who is alive and

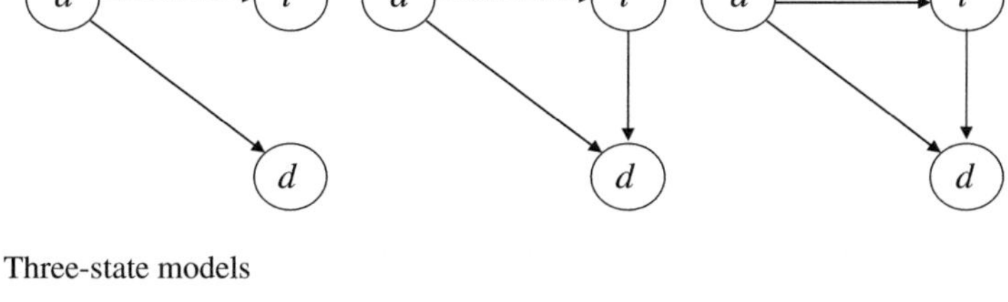

Fig. 6.2 Three-state models

not-permanently (or not-totally) disabled. Premiums are paid while the contract is in state a. The lump sum benefit is paid if disablement, i.e. transition $a \rightarrow i$, occurs. This insurance cover requires a probabilistic structure consisting of the probability of disablement (i.e. the probability of becoming disabled), and the probability of death of an active insured, both usually depending on the insured's attained age.

The model described above is very simple but rather unrealistic. It is more realistic to assume that the benefit will be paid out after a qualification period (see Sects. 3.2.4 and 6.7 devoted to the formal representation of policy conditions), which is required by the insurer in order to ascertain the permanent character of the disability. Similar simplifications will nevertheless be assumed in what follows, mainly focussing on the basic concepts.

A more complicated structure than the three-state model in Fig. 6.2a is needed in order to represent an annuity benefit in case of permanent (and total) disability. In this case, the death of the disabled insured must be considered. The resulting graph is depicted in Fig. 6.2b. Such a model is also called a double-decrement model with a second-order decrement, i.e. transition $i \rightarrow d$. The annuity benefit is assumed to be paid while the insured is disabled. Premiums are paid while the insured is active. The probabilistic structure also requires the probability of death for a disabled insured, usually as a function of his/her attained aged. Hence the assumptions about mortality must concern both active and disabled insureds. A hypothesis about mortality of insured lives, more complicated (and realistic) than the one assuming dependence on the attained age only, will be described in the following (see Sect. 6.8). A more realistic (and general) setting would allow for policy conditions such as the deferred period or the waiting period.

Let us generalize the structure described above by considering an annuity benefit in case of (total) disability; thus the permanent character of the disability is not required. Hence, we have to consider the possibility of recovery. The resulting model is represented by Fig. 6.2c. This insurance product requires a probabilistic structure also including the probabilities of recovery (or "reactivation"), i.e. transition $i \rightarrow a$.

Let us turn to the definition of a biometric model, in order to express in probabilistic terms the evolution of a risk throughout time. For this purpose we refer to the graph in Fig. 6.2c which represents a rather general structure. Actually, the probabilistic model for the graph in Fig. 6.2b can be obtained by assuming that the probability of transition $i \rightarrow a$ is equal to zero.

6.3.2 One-Year Transition Probabilities

A time-discrete approach is adopted, focussing in particular on a one-year time period. The probability that an individual aged y is in a certain state at age $y + 1$, conditional on being in a given state at age y, is called a (*one-year*) *transition probability*.

We assume that no more than one transition can occur during one year, apart from the possible death of the insured. This hypothesis is rather unrealistic when a time period of one year is concerned, as it disregards the possibility of short-lasting claims. It becomes more realistic if a smaller time unit is assumed. Of course, if the time unit is very small a time-continuous approach can be adopted (see Sect. 6.9, and the calculation methods described in Sects. 6.10–6.15).

Example 6.3.1 Table 6.1 shows some examples of transitions allowed by the assumption that no more than one transition can occur during one year (apart from possible death), and other transitions which are not allowed. □

We define the following one-year transition probabilities related to an active insured age y:

p_y^{aa} = probability of being active at age $y+1$
q_y^{aa} = probability of dying within one year, death occurring in state a
p_y^{ai} = probability of being disabled at age $y+1$
q_y^{ai} = probability of dying within one year, death occurring in state i

Further, we define the following one-year probabilities:

p_y^{a} = probability of being alive at age $y+1$
q_y^{a} = probability of dying within one year
w_y = probability of becoming disabled within one year

We recall that all the probabilities defined above are conditional probabilities, the conditioning event being the presence in state a at age y.

The following relations hold:

$$p_y^{aa} + p_y^{ai} = p_y^{a}, \tag{6.3.1}$$

$$q_y^{aa} + q_y^{ai} = q_y^{a}, \tag{6.3.2}$$

$$p_y^{a} + q_y^{a} = 1, \tag{6.3.3}$$

$$p_y^{ai} + q_y^{ai} = w_y. \tag{6.3.4}$$

Table 6.1 Examples of transitions between states

State at age y	Transition(s)	State at age $y+1$	Allowed by the model?
a	$\rightarrow$	i	Yes
i	$\rightarrow$	a	Yes
a	$\rightarrow i \rightarrow$	a	No
a	$\rightarrow i \rightarrow a \rightarrow$	i	No
a	$\rightarrow$	d	Yes
i	$\rightarrow$	d	Yes
a	$\rightarrow i \rightarrow$	d	Yes
i	$\rightarrow a \rightarrow$	d	Yes
a	$\rightarrow i \rightarrow a \rightarrow$	d	No

Now, let us consider a disabled insured aged y. We define the following one-year transition probabilities:

p_y^{ii} = probability of being disabled at age $y + 1$
q_y^{ii} = probability of dying within one year, death occurring in state i
p_y^{ia} = probability of being active at age $y + 1$
q_y^{ia} = probability of dying within one year, death occurring in state a

Moreover, we define the following one-year probabilities:

p_y^{i} = probability of being alive at age $y + 1$
q_y^{i} = probability of dying within one year
r_y = probability of recovery within one year

All the probabilities we have now defined are conditional probabilities, the conditioning event being the presence in state i at age y.

The following relations hold:

$$p_y^{ia} + p_y^{ii} = p_y^{i}, \tag{6.3.5}$$

$$q_y^{ii} + q_y^{ia} = q_y^{i}, \tag{6.3.6}$$

$$p_y^{i} + q_y^{i} = 1, \tag{6.3.7}$$

$$p_y^{ia} + q_y^{ia} = r_y. \tag{6.3.8}$$

Thanks to the assumption that no more than one transition can occur during one year (apart from possible death), probabilities p_y^{aa} and p_y^{ii} actually represent probabilities of remaining active and disabled respectively, from age y to $y + 1$.

When only permanent disability is addressed, we obviously have:

$$p_y^{ia} = q_y^{ia} = 0. \tag{6.3.9}$$

The conditional probabilities of being in one of the three states, a, i, d, at age $y + 1$, given the state occupied at age y, are represented in Table 6.2.

It is worth stressing that the only conditioning we have so far considered while defining the one-year probabilities is given by the state occupied by the insured at age y. For this reason, the probabilistic model we are constructing is a *Markov chain*. The set of probabilities represented in Table 6.2 constitutes the *stochastic matrix* (the

Table 6.2 Conditional probabilities of being in states a, i, d, at age $y + 1$

State at age y	State at age $y + 1$		
	a	i	d
a	p_y^{aa}	p_y^{ai}	q_y^{a}
i	p_y^{ia}	p_y^{ii}	q_y^{i}
d	0	0	1

sum of the items on each row is equal to 1), also called the *transition matrix*, related to the Markov chain.

More complex models, allowing for more detailed conditioning events, will be addressed in Sect. 6.8, whereas a formal description of Markov models will be provided in Sect. 6.9.

The set of probabilities needed for actuarial calculations can be reduced by adopting some *approximation formulae*. For example, common assumptions are as follows:

$$q_y^{ai} = w_y \frac{q_y^i}{2}, \tag{6.3.10}$$

$$q_y^{ia} = r_y \frac{q_y^a}{2}. \tag{6.3.11}$$

The hypotheses underpinning formulae (6.3.10) and (6.3.11) are as follows:

- uniform distribution of the first transition time within the year (the transition consisting in $a \to i$ or $i \to a$, respectively);
- the probability that the second transition ($i \to d$ or $a \to d$, respectively) occurs within the second half of the year is equal to one half of the probability that a transition of the same type occurs within the year.

More rigorous approximations can also be adopted; in particular:

$$q_y^{ai} = w_y \, {}_{\frac{1}{2}}q^i_{y+\frac{1}{2}}, \tag{6.3.12}$$

$$q_y^{ia} = r_y \, {}_{\frac{1}{2}}q^a_{y+\frac{1}{2}}, \tag{6.3.13}$$

with ${}_{\frac{1}{2}}q^i_{y+\frac{1}{2}}$ and ${}_{\frac{1}{2}}q^a_{y+\frac{1}{2}}$ derived from q_y^i and q_y^a respectively, via appropriate assumptions (e.g. uniform distribution of death within the year).

We note that, thanks to assumption (6.3.10), and assuming that the probabilities w_y, r_y, q_y^i and q_y^a (called *Zimmermann basic functions*) have been estimated from statistical data, all other probabilities can be calculated. In particular, we have:

$$p_y^{ai} = w_y - q_y^{ai} = w_y \left(1 - \frac{q_y^i}{2}\right), \tag{6.3.14}$$

$$p_y^{aa} = p_y^a - p_y^{ai} = p_y^a - w_y \left(1 - \frac{q_y^i}{2}\right), \tag{6.3.15}$$

$$q_y^{aa} = q_y^a - q_y^{ai} = q_y^a - w_y \frac{q_y^i}{2}. \tag{6.3.16}$$

6.3.3 Multi-year Transition Probabilities

The probabilities defined above refer to a one-year period. Of course, we can define probabilities relating to two or more years. The notation, for example referring to an active insured age y, is as follows:

${}_hp_y^{aa}$ = probability of being active at age $y+h$
${}_hp_y^{ai}$ = probability of being disabled at age $y+h$
etc.

Example 6.3.2 Consider a 4-year period. In Fig. 6.3 all the possible $2^4 = 16$ disability stories (i.e. "paths") are plotted, which start from state a at age y and terminate either in state a or i at age $y+4$. Assume we have to calculate the probability ${}_4p_y^{ai}$, given all the one-year probabilities p_{y+j}^{aa}, p_{y+j}^{ai}, p_{y+j}^{ia} and p_{y+j}^{ii}, for $j = 0, 1, 2, 3$. The state i is reached, at age $y+4$, by 8 of the 16 paths. Of course, the 8 stories are mutually exclusive, and hence the probability ${}_4p_y^{ai}$ is equal to the sum of the probabilities of the 8 stories. For example, the probability of the story

$$a \to a,\ a \to a,\ a \to a,\ a \to i$$

(see the path on the top of Fig. 6.3) can be expressed as follows, that is, as the product of one-year probabilities:

$$p_y^{aa}\, p_{y+1}^{aa}\, p_{y+2}^{aa}\, p_{y+3}^{ai}.$$

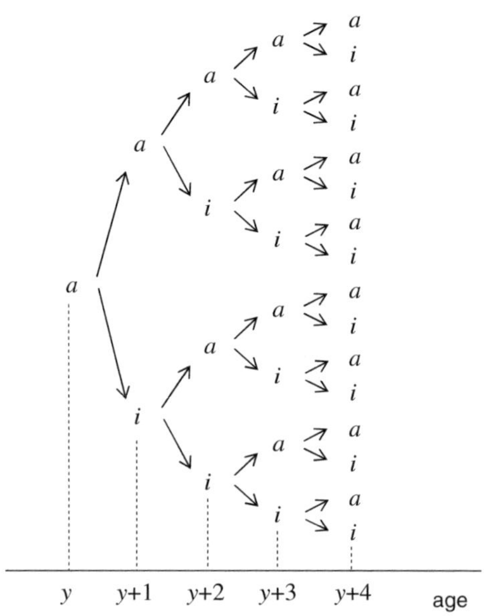

Fig. 6.3 Possible disability stories in a 4-year time interval

As is implicit in the notation, all the one-year probabilities take into account the current state only, disregarding the path which has lead to that state. Similarly, probabilities like ${}_kp_{y+j}^{ai}$ only rely on the state at age $y + j$. We note that, in order to eventually reach the state i, the final step must be either $a \to i$ or $i \to i$, depending on the state at age $y + 3$. Hence the probability ${}_4p_y^{ai}$ can be expressed as follows:

$$ {}_4p_y^{ai} = {}_3p_y^{aa}\, p_{y+3}^{ai} + {}_3p_y^{ai}\, p_{y+3}^{ii}, $$

where the probabilities ${}_3p_y^{aa}$ and ${}_3p_y^{ai}$ account for all the paths leading in 3 years from the state a at age y to the state a and i respectively at age $y + 3$. Probabilities ${}_3p_y^{aa}$ and ${}_3p_y^{ai}$ can, in turn, be expressed in a similar way. Generalizing leads to the recurrent relations (6.3.17) and (6.3.18). □

The following recurrent relationships (Chapman–Kolmogorov equations) involving one-year transition probabilities hold for $h \geq 1$:

$$ {}_hp_y^{aa} = {}_{h-1}p_y^{aa}\, p_{y+h-1}^{aa} + {}_{h-1}p_y^{ai}\, p_{y+h-1}^{ia}, \tag{6.3.17} $$

$$ {}_hp_y^{ai} = {}_{h-1}p_y^{aa}\, p_{y+h-1}^{ai} + {}_{h-1}p_y^{ai}\, p_{y+h-1}^{ii}, \tag{6.3.18} $$

with ${}_0p_y^{aa} = 1$ and ${}_0p_y^{ai} = 0$. For brevity, the relevant proof is omitted. Example 6.3.2 and Fig. 6.4 can help in interpreting the recursions.

The probabilities of remaining in a certain state for a given period are called *occupancy probabilities*. Let ${}_hp_y^{\overline{aa}}$, ${}_hp_y^{\overline{ii}}$ denote the probability for an individual aged y of remaining in state a, i respectively, for h years. Having excluded the possibility of more than one transition throughout the year, we first find, for $h = 1$:

$$ {}_1p_y^{\overline{aa}} = p_y^{aa}, \tag{6.3.19} $$

$$ {}_1p_y^{\overline{ii}} = p_y^{ii}. \tag{6.3.20} $$

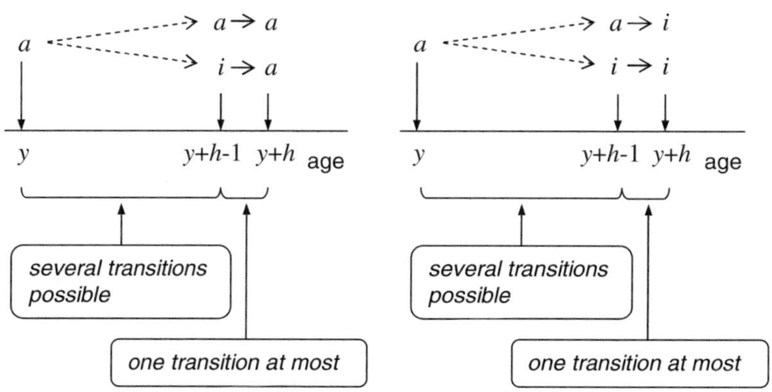

Fig. 6.4 Interpretation of recursions (6.3.17) and (6.3.18)

Then, the occupancy probabilities can in general be expressed, for $h \geq 1$, as follows:

$$ {}_h p_y^{\overline{aa}} = \prod_{k=0}^{h-1} p_{y+k}^{aa}, \tag{6.3.21} $$

$$ {}_h p_y^{\overline{ii}} = \prod_{k=0}^{h-1} p_{y+k}^{ii}. \tag{6.3.22} $$

Of course, ${}_0 p_y^{\overline{aa}} = {}_0 p_y^{\overline{ii}} = 1$.

Example 6.3.3 The set of paths which eventually lead to the state i at age $y+4$ can be split into the following four subsets (see Fig. 6.3):

1. paths entering the state i between age $y+3$ and $y+4$;
2. paths entering the state i between age $y+2$ and $y+3$, then remaining in i;
3. the path entering the state i between age $y+1$ and $y+2$, then remaining in i;
4. the path entering the state i between age y and $y+1$, then remaining in i.

Of course any sequence of states can be visited prior to the transition $a \to i$. Hence, the probability ${}_4 p_y^{ai}$ can be expressed as follows:

$$ {}_4 p_y^{ai} = \underbrace{{}_3 p_y^{aa}\, p_{y+3}^{ai}}_{\text{subset 1}} + \underbrace{{}_2 p_y^{aa}\, p_{y+2}^{ai}\, {}_1 p_{y+3}^{\overline{ii}}}_{\text{subset 2}} + \underbrace{{}_1 p_y^{aa}\, p_{y+1}^{ai}\, {}_2 p_{y+2}^{\overline{ii}}}_{\text{subset 3}} + \underbrace{p_y^{ai}\, {}_3 p_{y+1}^{\overline{ii}}}_{\text{subset 4}}. $$

□

Equation (6.3.18) leads to the following relationship, involving the probability of remaining disabled (for brevity, the relevant proof is omitted):

$$ {}_h p_y^{ai} = \sum_{r=1}^{h} \left[{}_{h-r} p_y^{aa}\, p_{y+h-r}^{ai}\, {}_{r-1} p_{y+h-r+1}^{\overline{ii}} \right]. \tag{6.3.23} $$

Equation (6.3.23) can be interpreted as follows (see also Example 6.3.3). Each term of the sum is the probability of a "story" which, starting from state a at age y, is in state a at age $y+h-r$ (probability ${}_{h-r} p_y^{aa}$), then in state i at age $y+h-r+1$ (probability p_{y+h-r}^{ai}), and finally remains in this state up to age $y+h$ (probability ${}_{r-1} p_{y+h-r+1}^{\overline{ii}}$). The probability, for an active insured aged y, of being disabled at age $y+h$ is then obtained by summing up the h terms. See also Fig. 6.5. As we will see, Eq. (6.3.23) has a central role in the calculation of actuarial values and premiums in particular.

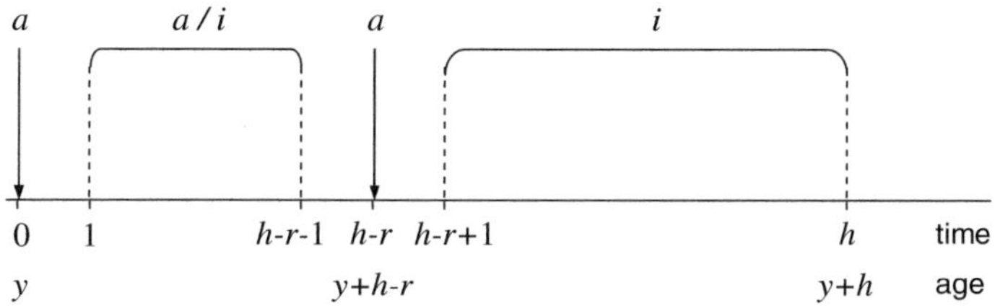

Fig. 6.5 Sequences of states and transitions: $a \to \cdots \to i$

6.4 Actuarial Values

The probabilities defined above allow us to express actuarial values (i.e. expected present values) concerning disability insurance. Hence, formulae for premiums calculated according to the equivalence principle follow.

Consider an individual active at age x (i.e. in state a). The actuarial value, $a^{ai}_{x:\overline{m}|}$, of the disability insurance cover described above (that is, the expected value of the random variable Y defined by (6.2.2), now with $b = 1$) is given by:

$$a^{ai}_{x:\overline{m}|} = \mathbb{E}[Y] = \sum_{h=1}^{m} v^h \; {}_h p^{ai}_x . \tag{6.4.1}$$

Using Eq. (6.3.23), we have:

$$a^{ai}_{x:\overline{m}|} = \sum_{h=1}^{m} v^h \sum_{r=1}^{h} \left[{}_{h-r} p^{aa}_x \; p^{ai}_{x+h-r} \; {}_{r-1} p^{\overline{ii}}_{x+h-r+1} \right] . \tag{6.4.2}$$

Then, letting $j = h - r + 1$ and inverting the order of summation in (6.4.2), we find:

$$a^{ai}_{x:\overline{m}|} = \sum_{j=1}^{m} {}_{j-1} p^{aa}_x \; p^{ai}_{x+j-1} \sum_{h=j}^{m} v^h \; {}_{h-j} p^{\overline{ii}}_{x+j} . \tag{6.4.3}$$

The quantity

$$\ddot{a}^{i}_{x+j:\overline{m-j+1}|} = \sum_{h=j}^{m} v^{h-j} \; {}_{h-j} p^{\overline{ii}}_{x+j} \tag{6.4.4}$$

is the actuarial value of a temporary immediate annuity paid to a disabled insured aged $x + j$ while he/she stays in state i (consistently with the definition of the insurance cover, the annuity is assumed to be payable up to the end of the policy term m), briefly the actuarial value of a *disability annuity*. By using (6.4.4), we finally obtain:

$$a^{ai}_{x:m\rceil} = \sum_{j=1}^{m} {}_{j-1}p^{aa}_{x}\; p^{ai}_{x+j-1}\; v^{j}\; \ddot{a}^{i}_{x+j:m-j+1\rceil}. \tag{6.4.5}$$

The right-hand side of Eq. (6.4.5) is an *inception-annuity* formula for the actuarial value of a disability annuity benefit. Indeed, it is based on the probabilities p^{ai}_{x+j-1} of entering state i and thus *becoming* disabled ("inception"), and the expected present value $\ddot{a}^{i}_{x+j:m-j+1\rceil}$ of an annuity payable while the insured *remains* disabled. Conversely, formula (6.4.1) expresses the same actuarial value in terms of the probabilities of *being* disabled at age $x + h$.

As regards the maximum benefit period we note the following.

- The preceding formulae are based on the assumption that the disability annuity is payable up to the policy term m. Hence the stopping time coincides with the policy term, while the maximum benefit period depends on the time at which the disability annuity starts.
- Conversely, policy conditions can state a maximum benefit period of s years, independent of the time at which the disability annuity starts. If s is large (compared to m) the benefit payment may last well beyond the insured period.

In Fig. 6.6 it is assumed that each individual disability "story" can be represented according to the Lexis diagram logic; j', j'', j''' denote three possible times at which the payment of the disability annuity starts, and the 45°-slope segments represent the related maximum benefit period. In particular, Fig. 6.6a refers to a disability annuity payable up to the policy term, whereas Fig. 6.6b refers to a disability annuity with a fixed maximum benefit period.

If a maximum benefit period of s years is stated, the actuarial value, which we denote by $a^{ai}_{x:m;s\rceil}$, can be easily derived from the right-hand side of Eq. (6.4.3):

$$a^{ai}_{x:m;s\rceil} = \sum_{j=1}^{m} {}_{j-1}p^{aa}_{x}\; p^{ai}_{x+j-1} \sum_{h=j}^{j+s-1} v^{h}\; {}_{h-j}p^{ii}_{x+j} \tag{6.4.6}$$

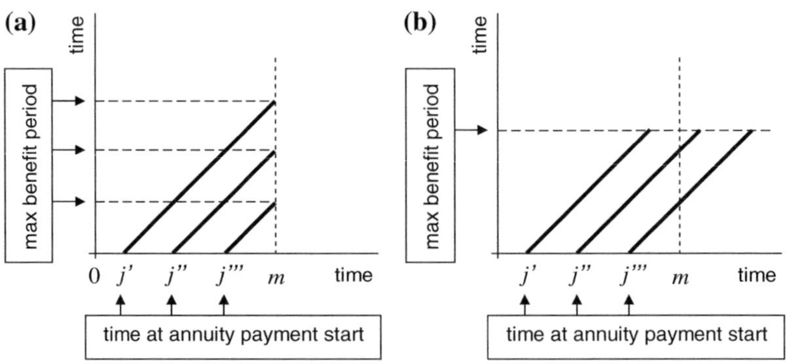

Fig. 6.6 Maximum benefit period according to policy conditions

or directly from Eq. (6.4.5) by changing the maximum duration of the disability annuity:

$$a^{ai}_{x:m;s\rceil} = \sum_{j=1}^{m} {}_{j-1}p^{aa}_{x}\ p^{ai}_{x+j-1}\ v^{j}\ \ddot{a}^{i}_{x+j:s\rceil}. \tag{6.4.7}$$

We note that a fixed benefit period s (or similar policy conditions, which will be described in Sect. 6.7) is required if the insured period m is very short, otherwise the insurance cover loses any practical meaning. In particular, a very short period, commonly $m = 1$, is stated for each individual cover in a group insurance policy (e.g. within an employee benefit plan) sponsored by the employer.

If $m = 1$, we obviously find:

$$a^{ai}_{x:1;s\rceil} = p^{ai}_{x}\ v\ \ddot{a}^{i}_{x+1:s\rceil}. \tag{6.4.8}$$

With reference to Eqs. (6.4.5) and (6.4.7) respectively, the quantities

$$p^{ai}_{x+j-1}\ v\ \ddot{a}^{i}_{x+j:m-j+1\rceil},$$
$$p^{ai}_{x+j-1}\ v\ \ddot{a}^{i}_{x+j:s\rceil}$$

represent, for $j = 1, 2, \ldots, m$, the annual expected costs to the insurer (for a unitary annual benefit) related to an active insured age $x + j - 1$. According to the traditional actuarial language, the annual expected costs are called the *natural premiums* of the insurance covers. The behavior of the natural premiums as functions of time (and hence age) should be carefully considered while checking the feasibility of premium arrangements (see Sect. 6.5).

According to (6.4.8), we can write:

$$p^{ai}_{x+j-1}\ v\ \ddot{a}^{i}_{x+j:m-j+1\rceil} = a^{ai}_{x+j-1:1;m-j+1\rceil}, \tag{6.4.9}$$

$$p^{ai}_{x+j-1}\ v\ \ddot{a}^{i}_{x+j:s\rceil} = a^{ai}_{x+j-1:1;s\rceil}. \tag{6.4.10}$$

Then, actuarial values (6.4.5) and (6.4.7) can also be expressed as expected present values (related to an active insured age x) of the annual expected costs. Indeed, we can write:

$$a^{ai}_{x:m\rceil} = \sum_{j=1}^{m} {}_{j-1}p^{aa}_{x}\ v^{j-1}\ a^{ai}_{x+j-1:1;m-j+1\rceil}, \tag{6.4.11}$$

$$a^{ai}_{x:m;s\rceil} = \sum_{j=1}^{m} {}_{j-1}p^{aa}_{x}\ v^{j-1}\ a^{ai}_{x+j:1;s\rceil}. \tag{6.4.12}$$

Let us consider a temporary immediate annuity payable for at most m' years while the insured (assumed to be active at age x) is active. The relevant actuarial value,

$\ddot{a}^{aa}_{x:m'\rceil}$, is given by

$$\ddot{a}^{aa}_{x:m'\rceil} = \sum_{h=1}^{m'} v^{h-1}\,{}_{h-1}p^{aa}_x\,. \tag{6.4.13}$$

This actuarial value is used for calculating periodic level premiums (if we assume that, as usual, premiums are waived during the disability spells).

6.5 Premiums

In this section we first focus on *net premiums*, i.e. premiums only meeting the benefits (and thus not accounting for insurer's expenses). As the premium calculation principle, we assume the *equivalence principle*. By definition, the equivalence principle is fulfilled if and only if, at the policy issue, the actuarial value of premiums is equal to the actuarial value of benefits.

The *net single premium* for an annual benefit b is then given by:

$$\Pi_{x,m} = b\, a^{ai}_{x:m\rceil} \tag{6.5.1}$$

if the disability benefit is payable up to the policy term m, and by:

$$\Pi_{x,m;s} = b\, a^{ai}_{x:m;s\rceil} \tag{6.5.2}$$

if a maximum benefit period of s years is stated.

Remark 1 As already noted in Sect. 5.2.2., the equivalence principle seems to be in contrast with a reasonable profit target. However, the equivalence principle can be "implemented" by adopting appropriate transition probabilities and interest rate (or discount factor), briefly a *safe-side technical basis* (or *first-order basis*), which originates a positive expected profit and, at the same time, lowers the probability of losses. We recall that the expression *second-order basis* conversely denotes a realistic description of the scenario which, in the context of disability insurance, consists of transition probabilities expressing a reliable assessment of disability, recovery and mortality, and the estimated yield from insurer's investment.

As regards, in particular, the mortality assumptions, it is worth noting that disability benefits are *living benefits*: benefits are indeed payable as long as the insured is alive (and disabled as well). Thus, a safe-side assessment of the insurer's liabilities related to insurance products providing disability annuities requires that the mortality of the insureds should not be overestimated. ■

When *periodic premiums* are involved, it is quite natural to assume that premiums are paid while the insured is active and not while disabled; indeed, premiums are usually waived during disability spells. Let us focus on annual level premiums payable for m' years at most ($m' \leq m$). If the annual benefit b is payable up to the policy term m, the equivalence principle is fulfilled if the annual premium, $P_{x,m(m')}$, satisfies the following equation:

$$P_{x,m(m')}\, \ddot{a}^{aa}_{x:\overline{m'}|} = b\, a^{ai}_{x:\overline{m}|}. \tag{6.5.3}$$

If a maximum benefit period of s years is stated, the premium $P_{x,m(m');s}$ is the solution of the following equation:

$$P_{x,m(m');s}\, \ddot{a}^{aa}_{x:\overline{m'}|} = b\, a^{ai}_{x:m;\overline{s}|}. \tag{6.5.4}$$

Let us assume $m' = m$. From Eqs. (6.5.3) and (6.4.13), and using (6.4.11), we find:

$$P_{x,m(m)} = b\, \frac{\sum_{j=1}^{m} {}_{j-1}p^{aa}_{x}\, v^{j-1}\, a^{ai}_{x+j-1:1;\overline{m-j+1}|}}{\sum_{j=1}^{m} v^{j-1}\, {}_{j-1}p^{aa}_{x}}. \tag{6.5.5}$$

Similarly, from Eqs. (6.5.4) and (6.4.13), and using (6.4.12), we obtain:

$$P_{x,m(m);s} = b\, \frac{\sum_{j=1}^{m} {}_{j-1}p^{aa}_{x}\, v^{j-1}\, a^{ai}_{x+j-1:1;\overline{s}|}}{\sum_{j=1}^{m} v^{j-1}\, {}_{j-1}p^{aa}_{x}}. \tag{6.5.6}$$

Then, in both cases, the annual level premium is an arithmetic weighted average of the natural premiums.

If natural premiums decrease as the duration of the policy increases, the level premium is initially lower than the natural premiums, leading to an insufficient funding of the insurer (resulting in a negative reserve). Consider Eq. (6.4.9), which expresses the natural premium $a^{ai}_{x+j-1:1;\overline{m-j+1}|}$. Although it is sensible to assume that the probability p^{ai}_{x+j-1} increases as the attained age $x + j - 1$ increases, the actuarial value $\ddot{a}^{i}_{x+j:\overline{m-j+1}|}$ may decrease because of the decreasing maximum benefit period, and hence the decreasing expected duration of the annuity payment. In this case, for a given insured period of m years, the number m' of premiums must be less than m. Conversely, this problem does not arise when the disability annuity is payable for a given number of years s.

Remark 2 Note, however, that the actuarial values $\ddot{a}^{i}_{x+j:\overline{m-j+1}|}$ and $\ddot{a}^{i}_{x+j:\overline{s}|}$ are also affected by the probability of recovery and the mortality of disabled people, which are not independent of age. ■

Example 6.5.1 We refer to an insurance product providing a disability annuity in the case of permanent or temporary disability; the annuity is payable at policy anniversaries, while the insured is disabled, and up to maturity m at most. Let $b = 100$ and $v = 1.02^{-1}$.

We assume:

$$p_y^{ai} = 0.00223 \times 1.0468^y.$$

We find, for example:

$$p_{30}^{ai} = 0.008795, \quad p_{45}^{ai} = 0.017465, \quad p_{55}^{ai} = 0.027594, \quad p_{60}^{ai} = 0.034684.$$

Further we assume:

$$p_y^{ia} = \begin{cases} 0.05 & \text{for } y \le 60, \\ 0 & \text{for } y > 60. \end{cases}$$

Let q_y denote the probability of dying between age y and $y+1$, irrespective of the state, and assume it is given by the Heligman–Pollard law, with parameters as specified in Sect. 2.3. As regards the state-specific mortality, we assume:

$$\begin{aligned} q_y^a &= q_y, \\ q_y^i &= (1+\gamma)\, q_y, \end{aligned}$$

with $\gamma = 0.25$. Finally, we have:

$$\begin{aligned} p_y^{aa} &= 1 - p_y^{ai} - q_y^a, \\ p_y^{ii} &= 1 - p_y^{ia} - q_y^i. \end{aligned}$$

Table 6.3 shows the single premiums and the annual level premiums for various ages at entry x and various policy terms m. For the same ages and policy terms, Table 6.4 shows annual level premiums payable for durations shorter than the policy term. Finally, Figs. 6.7 and 6.8 illustrate the behavior of natural premiums, related to various ages at entry, for a disability cover with $m = 10$ and $m = 20$ respectively. When natural premiums are decreasing, a level premium arrangement requires shortening the premium payment duration m' with respect to the policy duration m, in order to fulfill the financing condition. □

Table 6.3 Single premiums and annual level premiums ($m' = m$)

	$x = 30$	$x = 40$	$x = 50$
$m = 10$	41.656	64.219	96.918
	4.756	7.541	11.945
$m = 15$	84.185	127.360	190.563
	6.919	10.965	17.758
$m = 20$	136.777	202.044	311.067
	9.129	14.411	24.816

Table 6.4 Annual level premiums

	$x = 30$	$x = 40$	$x = 50$
$m = 10$ $m' = 7$	6.495	10.197	15.876
$m = 15$ $m' = 10$	9.611	14.956	23.487
$m = 20$ $m' = 15$	11.242	17.395	28.988

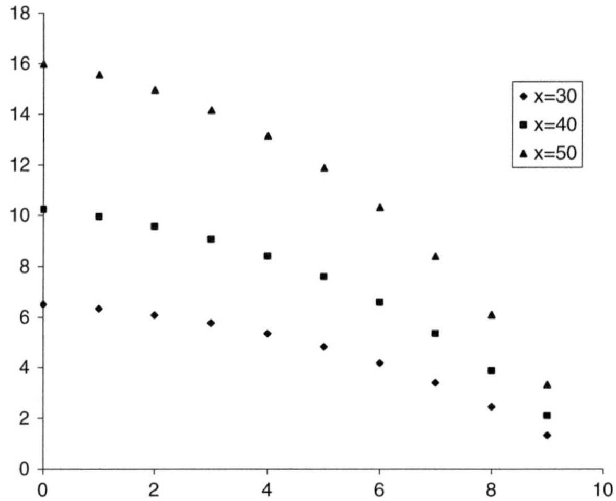

Fig. 6.7 Natural premiums; $m = 10$

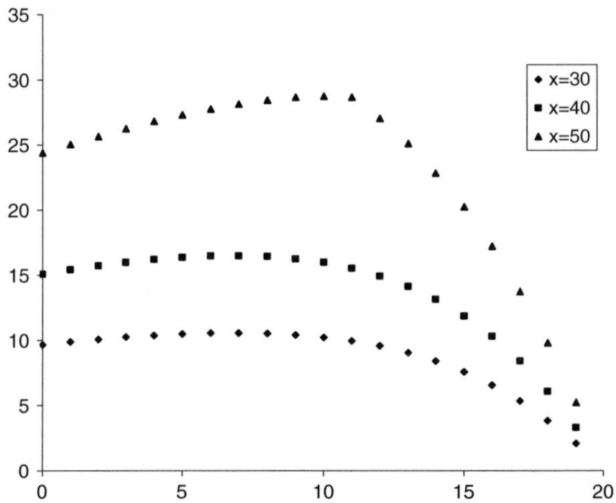

Fig. 6.8 Natural premiums; $m = 20$

Premiums paid by the policyholders are *gross premiums* (also called *office premiums*). Gross premiums are determined from equivalence premiums by adding the expense loading, and, possibly, a profit/safety loading (especially when the technical basis is not chosen according to a prudential principle; see also Sect. 5.2.2). It is worth noting that the expense structure, also for disability insurance covers, is more complicated than in life insurance. Expenses are also incurred in the event of a claim, and result from both the initial investigations in assessing claims and the periodic investigations needed to control the continuation of claims.

6.6 Reserves

As is well known, in actuarial mathematics the (prospective) policy reserve at time t is defined as the actuarial value of future benefits less the actuarial value of future premiums, given the state of the policy at time t. Thus, in disability insurance we have to define an *active reserve* (related to state a) as well as a *disabled reserve* (state i).

6.6.1 Active and Disabled Reserves

For brevity, let us only address the disability cover whose actuarial value is given by Eq. (6.4.5). Level premiums are payable for m' years. For simplicity, let $P = P_{x,m(m')}$.

The active reserve at (integer) time t is then given by:

$$V_t^{(a)} = \begin{cases} b\, a^{ai}_{x+t:m-t\rceil} - P\, \ddot{a}^{aa}_{x+t:m'-t\rceil} & 0 \le t < m', \\ b\, a^{ai}_{x+t:m-t\rceil} & m' \le t \le m, \end{cases} \tag{6.6.1}$$

with, of course, $V_0^{(a)} = V_m^{(a)} = 0$.

Conversely, the disabled reserve is given by the following expression:

$$V_t^{(i)} = \begin{cases} b\, \ddot{a}^{ii}_{x+t:m-t\rceil} - P\, \ddot{a}^{ia}_{x+t:m'-t\rceil} & 0 \le t < m', \\ b\, \ddot{a}^{ii}_{x+t:m-t\rceil} & m' \le t \le m. \end{cases} \tag{6.6.2}$$

Note, in particular, that the term $b\, \ddot{a}^{ii}_{x+t:m-t\rceil}$ on the right-hand side of Eq. (6.6.2) (on both the first and the second line) expresses the actuarial value of the running disability annuity as well as of possible future disability annuities after recovery. The term $P\, \ddot{a}^{ia}_{x+t:m'-t\rceil}$ is the actuarial value of future premiums paid after possible recovery.

After several manipulations, we find the following recursive relations, for the active reserve and the disabled reserve respectively:

$$V_t^{(a)} + P = v\,V_{t+1}^{(a)} + v\,p_{x+t}^{ai}\left(V_{t+1}^{(i)} - V_{t+1}^{(a)}\right) - v\,q_{x+t}^{a}\,V_{t+1}^{(a)}, \tag{6.6.3}$$

$$V_t^{(i)} - b = v\,V_{t+1}^{(i)} + v\,p_{x+t}^{ia}\left(V_{t+1}^{(a)} - V_{t+1}^{(i)}\right) - v\,q_{x+t}^{i}\,V_{t+1}^{(i)}. \tag{6.6.4}$$

The interpretation of Eqs. (6.6.3) and (6.6.4) is straightforward. Nevertheless, it is worth stressing the meaning of two specific items. The quantity $\left(V_{t+1}^{(i)} - V_{t+1}^{(a)}\right)$ in Eq. (6.6.3) represents the jump in the reserve required because of the transition $a \to i$, i.e. from the active to the disabled state. As shown in Example 6.6.1, the amount of the disabled reserve is much larger than that of the active reserve. Hence, the (positive) jump in the reserve profile must be financed via a mutuality mechanism, that is, allocating to the reserve of the insured entering the disability state a share of the premiums paid by other policyholders belonging to the portfolio. This mechanism is perfectly similar to the one which underpins the payment of the death benefit, for example, in a term insurance portfolio. See Fig. 6.9; note that the graph only provides an idea of the reserve profile, as a realistic relation between the size of the active and the disabled reserve is not respected.

Conversely, the quantity $\left(V_{t+1}^{(a)} - V_{t+1}^{(i)}\right)$ in Eq. (6.6.4) represents the (negative) jump corresponding to the insured's recovery, that is, transition $i \to a$. This implies a release of a large part of the policy reserve, which is shared among the other policies in the portfolio.

In actuarial practice, however, the calculation of the disabled reserve is often simplified by adopting some reasonable approximation formulae. In particular, we note that Eq. (6.6.2) can be reinterpreted as follows. The term $b\,\ddot{a}_{x+t:\overline{m-t}|}^{ii}$ includes both the actuarial value of the running disability annuity and the actuarial value of future possible disability annuities (after recovery) as well. Similarly, if $t < m'$, the term $P\,\ddot{a}_{x+t:\overline{m'-t}|}^{ia}$ represents the actuarial value of premium payable in possible future intervals spent in the active state. Hence, the disabled reserve can be split into two terms:

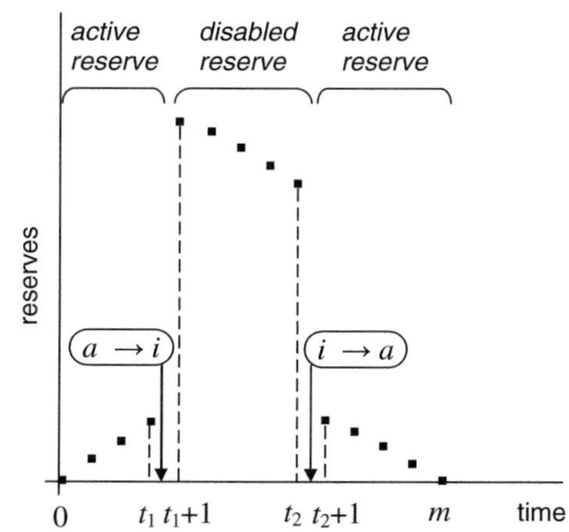

Fig. 6.9 A time-profile of the policy reserve

1. a term equal to the actuarial value of the running disability annuity, relating to the current disability spell;
2. a term equal to the actuarial value of benefits relating to future disability spells less the actuarial value of premiums payable after recovery.

If we neglect term 2, then we obtain a simpler (approximate) expression for the disabled reserve:

$$V_t^{(i)} = b\,\ddot{a}^{i}_{x+t:m-t\rceil}. \tag{6.6.5}$$

Of course, no approximation is involved by Eq. (6.6.5) if we refer to a disability annuity only payable in the case of permanent disability.

Remark As regards the choice of the probabilistic basis in premiums and reserve calculation, a problem arises, similar to the one discussed in Sect. 5.3. The choice depends on the transferability of the active reserve in the case the policyholder stops premium payment and withdraws from the contract (of course, assuming that the policyholder may also withdraw during disability spells does not reflect a likely situation). If the (active) reserve is transferable, the usual survival probability is required in the calculations. On the contrary, if the reserve is retained by the insurer, the probability that the policy is in-force is needed, as both mortality and lapses must be accounted for. ■

Example 6.6.1 Assume the technical basis adopted in Example 6.5.1. Figures 6.10 and 6.11 show the active reserve as a function of time t elapsed since policy issue, for disability covers with $m = 10$ and $m = 20$, respectively. In both cases, it clearly

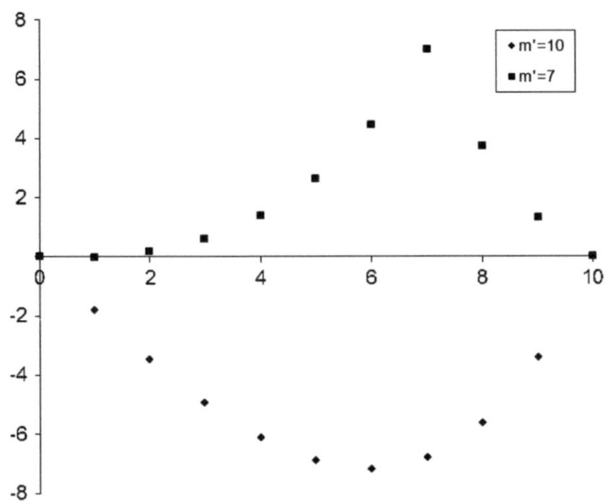

Fig. 6.10 Active reserves; $x = 30$, $m = 10$

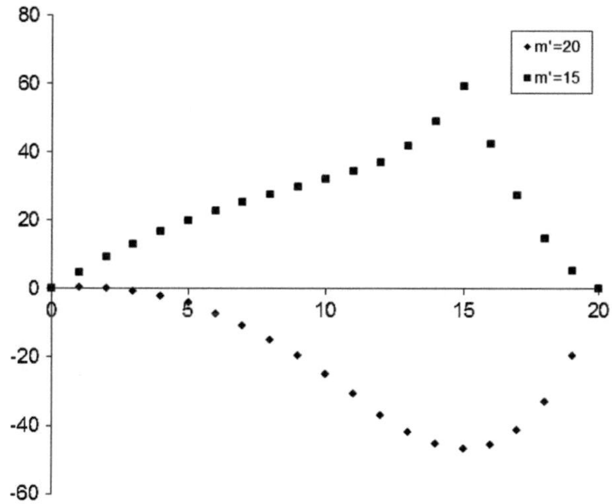

Fig. 6.11 Active reserves; $x = 50, m = 20$

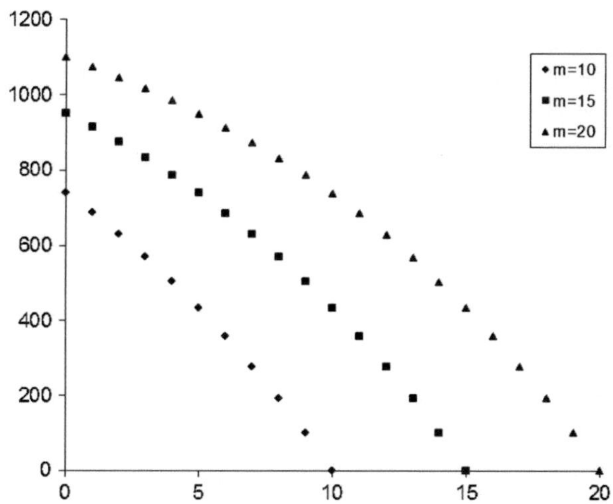

Fig. 6.12 Disabled reserves; $x = 30$

emerges that premium payment must be shortened with respect to policy duration in order to avoid negative reserves (which would represent an insurer's credit), and hence to fulfill the financing condition. Figure 6.12 shows the behavior of the disabled reserve, calculated by adopting the simplified formula (6.6.5). In particular, it is interesting to note the much higher amounts involved in the disabled reserve, when compared to the amounts related to active reserves. Indeed, the disabled reserve refers to an annuity in payment (see Eq. (6.6.5)), whereas the active reserve accounts for possible future disability annuities according to (6.6.1). □

The presence of policy conditions such as a deferred period or a waiting period leads to more complicated expressions for the reserves. We just mention a particular reserving problem arising from the presence of a deferred period.

Let us refer to a disability group insurance written on, say, a one-year basis and assume a deferred period of f. At the end of the one-year risk period, the insurer will need to set up reserves in respect of:

1. lives who are currently claiming;
2. lives who are currently disabled but whose period of disability has not yet reached the deferred period f and so are not currently claiming.

The reserve for category 2 is an *IBNR-type reserve*, i.e. a reserve for Incurred But Not Reported claims, widely discussed in the non-life insurance literature.

6.6.2 Reserves at Fractional Durations

When working in a time-discrete framework (as we are actually doing), we can obtain approximations to the exact value of the active reserve and the disabled reserve via interpolation procedures, in particular by adopting linear interpolation formulae.

As regards the active reserve, the reader can refer to formulae proposed in Sect. 5.3.4; see, in particular, Eqs. (5.3.15) and (5.3.16), and Figs. 5.7 and 5.8.

The disabled reserve can be assessed, for $t = 0, 1, \dots$ and $0 < r < 1$, as follows:

$$V_{t+r}^{(i)} = (1 - r)\,(V_t^{(i)} - b) + r\,V_{t+1}^{(i)}. \tag{6.6.6}$$

The resulting interpolated profile is represented in Fig. 6.13.

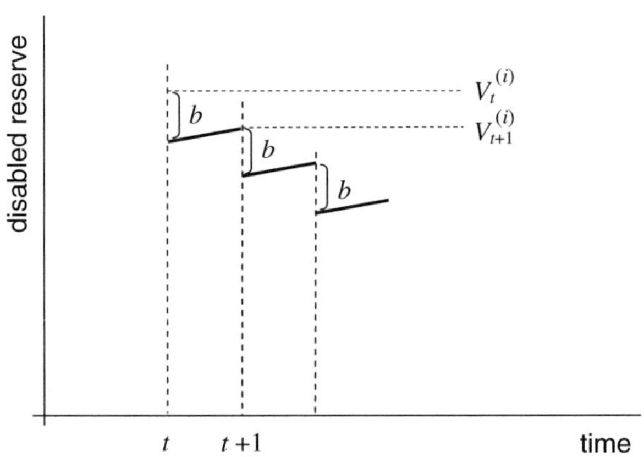

Fig. 6.13 Interpolated profile of the disabled reserve

6.7 Representing Policy Conditions

Several policy conditions, interesting insurance products providing a disability annuity, have been described in Sect. 3.2.4. In particular, conditions can refer to:

- the *insured period* (or *coverage period*): an annuity is payable if the disability inception time belongs to this interval;
- the duration of payment of the disability annuity.

The most important policy conditions can be formally described by a set of five parameters:

$$\Gamma = [m_1, m_2, f, s, r], \tag{6.7.1}$$

where

(m_1, m_2) = insured period; for example:

$m_1 = c$ = waiting period (from policy issue)
$m_2 = m$ = policy term

f = deferred period (from disability inception)
s = maximum benefit period, i.e. maximum number of years of annuity payment (from disability inception)
r = stopping time (from policy issue); for example, if x denotes the age at policy issue and ξ the retirement age, then we can set $r = \xi - x$

We again assume (as in Fig. 6.6) that each individual "story" can be represented according to the Lexis diagram logic. See Fig. 6.14, in which each horizontal segment represents a period during which the insured is active, while each 45°-slope segment represents a disability period. The set of all possible individual stories is then represented by the shaded region in Fig. 6.15.

The actuarial value, and hence the single premium, of an insurance product providing a disability annuity according to a given set Γ of policy conditions can

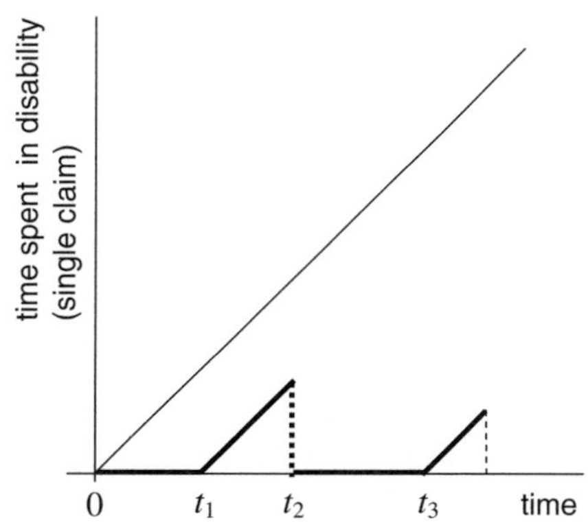

Fig. 6.14 An example of a disability story

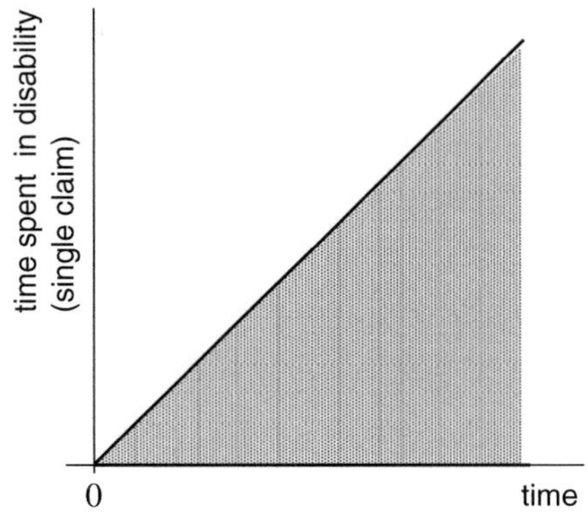

Fig. 6.15 The region of possible disability stories

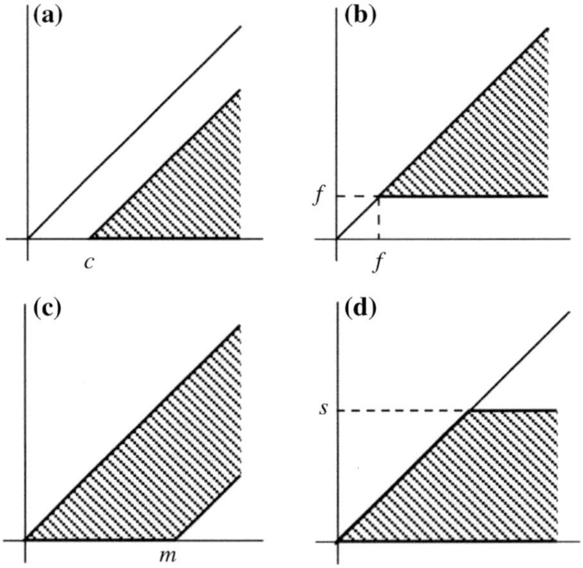

Fig. 6.16 Policy conditions, (**a**) $\Gamma = [c, \infty, 0, \infty, \infty]$, (**b**) $\Gamma = [0, \infty, f, \infty, \infty]$, (**c**) $\Gamma = [0, m, 0, \infty, \infty]$, (**d**) $\Gamma = [0, \infty, 0, s, \infty]$

be interpreted as a "measure" associated to a subset of the shaded region, the subset being determined by Γ.

Figures 6.16 and 6.17 show the subsets related to various policy conditions. Of course, several policy conditions can be combined together, resulting in more complicated structures.

Example 6.7.1 Consider the disability insurance product described in Sect. 6.4, according to which the disability annuity payment can last at most up to the policy term m. The actuarial value is given by Eq. (6.4.3). Policy conditions describing

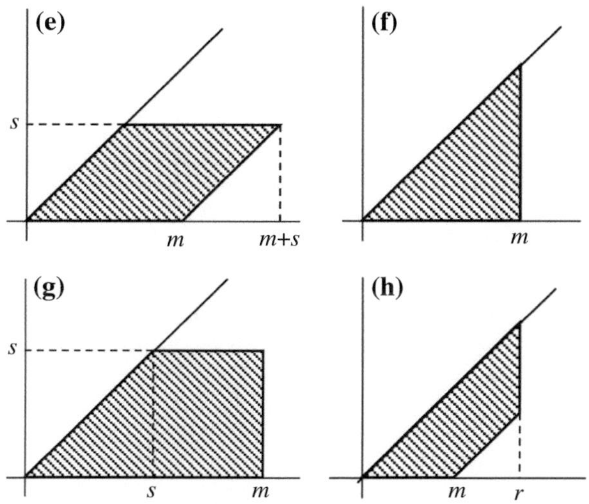

Fig. 6.17 Policy conditions, (**e**) $\Gamma = [0, m, 0, s, \infty]$, (**f**) $\Gamma = [0, m, 0, \infty, m]$, (**g**) $\Gamma = [0, m, 0, s, m]$, (**h**) $\Gamma = [0, m, 0, \infty, r]$

this basic structure are as follows:

$$\Gamma = [0, m, 0, \infty, m].$$

See Fig. 6.17f. It clearly follows that the "measure" associated to a specific subset of the region shown in Fig. 6.15 is a double summation (and a double integral in the time-continuous framework).

Consider the insurance product, also described in Sect. 6.4, according to which the maximum benefit period is equal to s years. The actuarial value is given by Eq. (6.4.6). Policy conditions are hence described by

$$\Gamma = [0, m, 0, s, \infty].$$

See Fig. 6.17e.

More complicated formulae are needed to allow for other policy conditions, e.g. the deferred period f. We do not deal with these aspects; some examples will be provided in Sect. 6.11, when dealing with practical calculation methods. □

6.8 Allowing for Duration Effects

The probabilistic model defined above (see Sect. 6.3) assumes that, for an insured aged x at policy issue, transition probabilities at any age y ($y \geq x$) only depend on the current state at that age, whereas they are independent of the health story before age y. More realistic (and possibly more complex) models can be built.

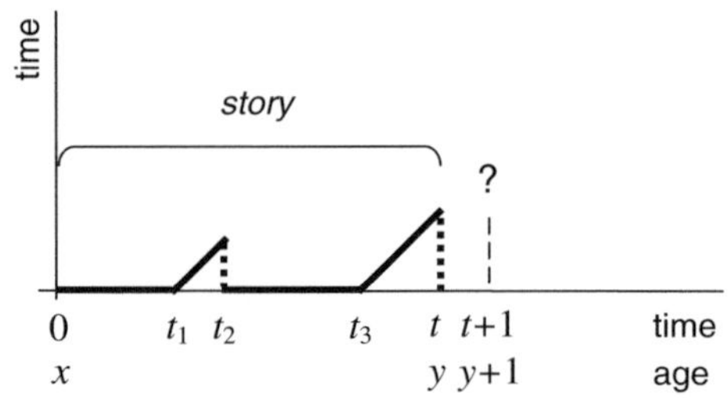

Fig. 6.18 The story of an insured risk

Consider Fig. 6.18. As regards the possible transition $i \to a$ between time t and $t+1$, we can consider the following dependence assumptions, expressed in terms of conditional probabilities:

1. $\mathbb{P}[\text{in } a \text{ at age } y+1 \mid \text{time elapsed since policy issue}]$
2. $\mathbb{P}[\text{in } a \text{ at age } y+1 \mid \text{story up to age } y]$
3. $\mathbb{P}[\text{in } a \text{ at age } y+1 \mid \text{total time in } i \text{ up to age } y]$
4. $\mathbb{P}[\text{in } a \text{ at age } y+1 \mid \text{time in } i \text{ since the latest transition into } i]$
5. $\mathbb{P}[\text{in } a \text{ at age } y+1 \mid \text{number of transitions into } i \text{ up to age } y]$
6. $\mathbb{P}[\text{in } a \text{ at age } y+1 \mid \text{in } i \text{ at age } y]$

Again referring to Fig. 6.18, we have:

1. time $= t$ (i.e. the past duration of the policy)
2. story $= (a, i, a, i;\ t_1, t_2, t_3)$
3. total time $= (t_2 - t_1) + (t - t_3)$
4. time $= (t - t_3)$ (i.e. the past duration of the current disability spell)
5. number $= 2$

First, we note that assumption 6, which leads to the *Markov model*, is the one we have adopted so far (see also Sect. 6.3.2). Its implementation is rather simple, and hence it is widely adopted in the actuarial practice.

The consideration of dependence 1, named *duration-since-initiation dependence*, implies the use of *issue-select* probabilities, i.e. probabilities which are functions of both x and t (rather than functions of the attained age $y = x + t$ only). For example, issue selection in the probability of transition $a \to i$ can represent a lower risk of disablement because of sickness thanks to a medical ascertainment carried out at the policy issue. In what follows we do not consider this aspect.

Assumption 2 implies serious difficulties in finding appropriate models which link the transition probability to any possible past story.

Assumptions 3, 4 and 5 lead to complex non-Markov models; however it is possible to shift to Markov models via approximations.

In particular, allowing for dependence 4, the *duration-in-current-state dependence*, requires *inception-select* probabilities depending on both the attained age $y = x + t$ and the time z spent in the current state ("inception" denoting the time at which the latest transition to that state occurred). Practical issues suggest that we focus on transitions from state i, i.e. on the disability duration effect on recovery and

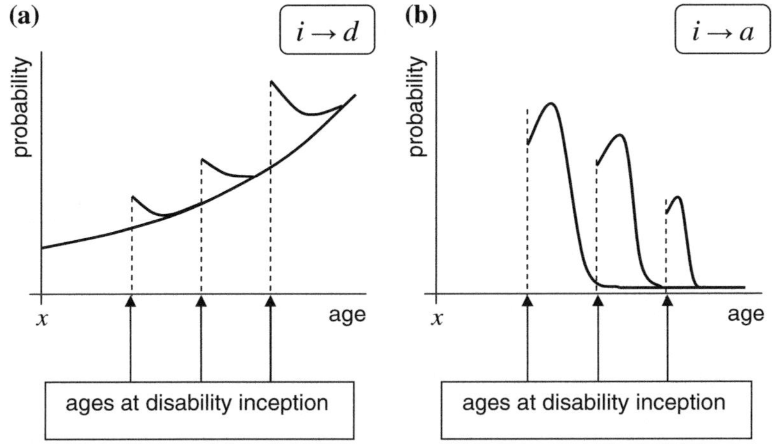

Fig. 6.19 Effect of time spent in the disability state (**a**) mortality of a disabled insured (**b**) recovery of a disabled insured

mortality of disabled lives. Actually, statistical evidence reveals an initial "acute" phase and then a "chronic" (albeit not necessarily permanent) phase of the disability spell. In the former both recovery and mortality have high probabilities, whilst in the latter recovery and mortality have lower probabilities. See Fig. 6.19.

Some examples of inception-select probabilities (according to model 4), related to a disabled insured, follow. Probabilities are denoted adopting the usual actuarial notation. We note that disability inception at age y must be interpreted as inception between the exact ages $y-1$ and y; hence, z represents the integer part of the actual past duration of the current disability spell.

$$p^{ia}_{[y]} = \mathbb{P}[\text{in } a \text{ at age } y+1 \mid \text{disability inception at age } y] \quad (\text{i.e. } z=0)$$
$$p^{ia}_{[y-z]+z} = \mathbb{P}[\text{in } a \text{ at age } y+1 \mid \text{disability inception at age } y-z]$$
$${}_k p^{\overline{ii}}_{[y]} = \mathbb{P}[\text{in } i \text{ up to age } y+k \mid \text{disability inception at age } y] \quad (\text{i.e. } z=0)$$

Similarly, the inception-select actuarial value of a disability annuity, payable to an individual at age y who entered the disability state at age $y-z$, is denoted as follows:

$\ddot{a}^{i}_{[y-z]+z:\overline{m-z+1}|}$ if the annuity is payable up to the policy term m at most;
$\ddot{a}^{i}_{[y-z]+z:\overline{s-z+1}|}$ if the annuity is payable for s years at most.

These actuarial values should in particular be used to calculate the disabled reserve. An example of inception-select actuarial value will be provided in Sect. 6.11.

Still focussing on dependence 4, it is worth noting that disability duration effects can be introduced in actuarial modelling without formally defining probabilities depending on both the attained age and the disability duration, and hence working with a Markov model. The key idea consists of splitting the disability state i into n states, $i^{(1)}, i^{(2)}, \dots, i^{(n)}$ (see Fig. 6.20), which represent disability according to duration since disablement.

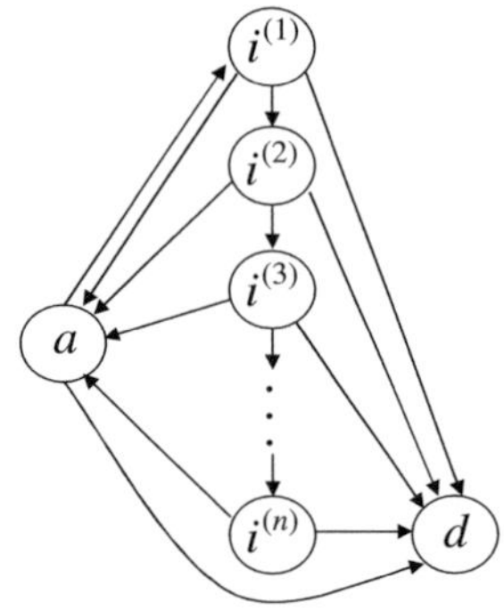

Fig. 6.20 The "Dutch" model

For example, the meaning of the n disability states can be as follows:

$i^{(h)}$ = the insured is disabled with a duration of disability between $h - 1$ and h, for $h = 1, 2, \ldots, n - 1$;
$i^{(n)}$ = the insured is disabled with a duration of disability greater than $n - 1$.

The conditional probabilities, according to this setting, constitute the stochastic matrix represented by Table 6.5.

The disability duration effect can be expressed via an appropriate choice of the involved probabilities. For instance, it is sensible to assume that for any age y:

$$p_y^{i^{(1)}a} > p_y^{i^{(2)}a} > \cdots > p_y^{i^{(n)}a} \geq 0. \tag{6.8.1}$$

A disability actuarial model allowing for duration dependence via splitting of the disability state has been adopted in the Netherlands, and is often called the *Dutch model*. In most applications it is assumed that $n = 6$.

Obviously, splitting the disability state leads to more complicated expressions for actuarial values and, in particular, premiums and reserves.

We finally consider a (simplified) implementation of assumption 5 via a Markov model. Define the following states (see Fig. 6.21):

Table 6.5 Conditional probabilities of being in states $a, i^{(1)}, i^{(2)}, \ldots, i^{(n)}, d$, at age $y + 1$

State at age y	State at age $y+1$					
	a	$i^{(1)}$	$i^{(2)}$	...	$i^{(n)}$	d
a	p_y^{aa}	$p_y^{ai^{(1)}}$	0	...	0	q_y^a
$i^{(1)}$	$p_y^{i^{(1)}a}$	0	$p_y^{i^{(1)}i^{(2)}}$	...	0	$q_y^{i^{(1)}}$
$i^{(2)}$	$p_y^{i^{(2)}a}$	0	0	...	0	$q_y^{i^{(2)}}$
...	...	...	...	...	...	...
$i^{(n)}$	$p_y^{i^{(n)}a}$	0	0	...	$p_y^{i^{(n)}i^{(n)}}$	$q_y^{i^{(n)}}$
d	0	0	0	...	0	1

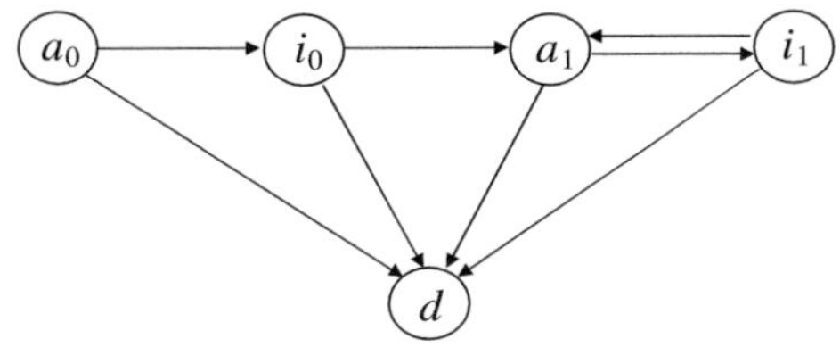

Fig. 6.21 A model with active and disabled states split according to previous disability

Table 6.6 Conditional probabilities of being in states a_0, i_0, a_1, i_1, d, at age $y+1$

State at age y	State at age $y+1$				
	a_0	i_0	a_1	i_1	d
a_0	$p_y^{a_0a_0}$	$p_y^{a_0i_0}$	0	0	$q_y^{a_0}$
i_0	0	$p_y^{i_0i_0}$	$p_y^{i_0a_1}$	0	$q_y^{i_0}$
a_1	0	0	$p_y^{a_1a_1}$	$p_y^{a_1i_1}$	$q_y^{a_1}$
i_1	0	0	$p_y^{i_1a_1}$	$p_y^{i_1i_1}$	$q_y^{i_1}$
d	0	0	0	0	1

a_0 = active, no previous disability
i_0 = disabled, no previous disability
a_1 = active, previously disabled
i_1 = disabled, previously disabled

The conditional probabilities, according to this setting, constitute the stochastic matrix represented by Table 6.6.

The rationale of splitting both the active state and disabled state is the assumption of a higher risk of disablement, a higher probability of death, and a lower probability of recovery for an insured who has already experienced disability. So, for example, we can reasonably assume that, for any age y:

$$p_y^{i_0a_1} > p_y^{i_1a_1}; \quad p_y^{a_1i_1} > p_y^{a_0i_0}. \tag{6.8.2}$$

Note that, also in this model, various probabilities do depend to some extent on the history of the insured risk before time t, but an appropriate definition of the states allows for this dependence in the framework of a Markov model.

6.9 Multistate Models: The Markov Assumption

A unifying approach to actuarial problems concerning a number of insurance covers within the area of insurances of the person (life insurance, disability covers, long term care products, etc.) can be achieved thanks to the mathematics of Markov stochastic processes, both in a time-continuous and a time-discrete context. The resulting models are usually called Markov multistate models.

In this section we do not deal in depth with mathematical features of Markov multistate models. A simple introduction will be provided, aiming at a more general approach to disability actuarial problems. In particular, we will see how Markov multistate models can help in understanding premium calculation methods commonly used in actuarial practice.

As seen in Sect. 6.3, the evolution of a risk (e.g. an insured individual) can be described in terms of the presence of the risk itself, at every point of time, in a certain *state* belonging to a given set of states, or *state space*. Formally, we denote by $\mathscr{S}$ the state space. We assume that $\mathscr{S}$ is a finite set. Referring to a disability insurance cover, the (simplest) state space is:

$$\mathscr{S} = \{a, i, d\}. \tag{6.9.1}$$

The set of direct *transitions* is denoted by $\mathscr{T}$. Let us denote each transition by an ordered pair. For example, (a, i) denotes the transition $a \rightarrow i$, i.e. the disablement. Referring to an insurance cover providing an annuity in case of (non-necessarily permanent) disability, we have:

$$\mathscr{T} = \{(a, i), (i, a), (a, d), (i, d)\}. \tag{6.9.2}$$

The pair $(\mathscr{S}, \mathscr{T})$ is called a *multistate model*. A graphical representation of a multistate model is provided by a *directed graph*; see, for instance, Fig. 6.2a–c. As already noted, a multistate model simply describes the contingencies pertaining to an insured risk. A probabilistic structure is needed in order to express a numerical assessment of the contingencies.

Let us suppose that we are at the policy issue, i.e. at time 0. The time unit is one year. Let $S(t)$ denote the random state occupied by the risk at time t, where, in a time-continuous context, t is a non-negative real number. Of course, $S(0)$ is a given state; in disability insurance, usually it is assumed that $S(0) = a$. The process $\{S(t); t \geq 0\}$ is a *time-continuous stochastic process*, with values in the finite set $\mathscr{S}$. The variable t is often called the *seniority*; it represents the past duration of the policy. If we refer to an insured age x at policy issue, then $x + t$ represents the attained age. Any possible realization $\{s(t)\}$ of the process $\{S(t)\}$ is called a *sample path*; thus, $s(t)$ is a function of the non-negative variable t, with values in $\mathscr{S}$.

Conversely, in a time-discrete context the variable t takes a finite number of values, in particular integer values; then, the *time-discrete stochastic process* $\{S(t);\ t = 0, 1, \dots\}$ is referred to. Note that this process has been implicitly assumed in the preceding sections while dealing with probabilities and actuarial values for disability insurance.

We now define a probabilistic structure for the stochastic process $\{S(t)\}$. First, let us refer to the time-discrete context. Assume that, for all integer times t, u, with $u > t \geq 0$, and for each pair of states j, k, the following property is satisfied:

$$\mathbb{P}[S(u) = k \mid S(t) = j \wedge H(t)] = \mathbb{P}[S(u) = k \mid S(t) = j], \tag{6.9.3}$$

where $H(t)$ denotes any hypothesis about the path $\{s(\tau)\}$ for $\tau < t$. Thus, it is assumed that the conditional probability on the left-hand side of Eq. (6.9.3) only depends on the "most recent" information $\{S(t) = j\}$ and is independent of the path before t. The process $\{S(t); t = 0, 1, \dots\}$ is then a *time-discrete Markov chain.*

Let us go back to transition probabilities already defined. For example, consider the probability denoted by p_y^{ai}, with $y = x + t$. The notation we have now defined yields:

$$p_{x+t}^{ai} = \mathbb{P}[S(t+1) = i \mid S(t) = a]. \tag{6.9.4}$$

Moreover, referring to the (more general) probability denoted by ${}_h p_y^{ai}$, we have:

$${}_h p_{x+t}^{ai} = \mathbb{P}[S(t+h) = i \mid S(t) = a]. \tag{6.9.5}$$

These equalities witness that the Markov assumption is actually adopted when defining the usual probabilistic structure for disability insurance. It should be stressed that, although an explicit use of multistate Markov models dates back to the end of the 1960s, the basic mathematics of what we now call a Markov chain model was developed during the eighteenth century and the first systematic approach to disability actuarial problems, consistent with the Markov assumption, dates back to the beginning of the 1900s (see Sect. 6.18 for relevant references).

An appropriate definition of the state space $\mathscr{S}$ allows us to express more general hypotheses of dependence, still remaining in the context of Markov chains. An important practical example is provided by the splitting of the disability state i into a set of states referring to various disability durations, as we have seen in Sect. 6.8. The resulting state space

$$\mathscr{S} = \{a, i^{(1)}, i^{(2)}, \dots, i^{(n)}, d\} \tag{6.9.6}$$

(represented in Fig. 6.20) allows for disability duration effects on recovery and mortality of disabled lives. Then, in discrete time, dependence on duration becomes dependence on the current state.

Another interesting example is provided by the multistate model represented in Fig. 6.21, in which the splitting is no longer based on durations but on the occurrence of (one or more) previous disability periods.

We now turn to a time-continuous context. While in a time-discrete approach the probabilistic structure is assigned via one-year transition probabilities (e.g. p_y^{aa}, p_y^{ai}, etc.), in a time-continuous setting it is usual to resort to *transition intensities* (or *forces* or *instantaneous rates*).

Let us use the following notation:

$$P^{jk}(t, u) = \mathbb{P}[S(u) = k \mid S(t) = j] \quad (j \neq k). \tag{6.9.7}$$

The transition intensity μ_t^{jk} is then defined as follows:

$$\mu_t^{jk} = \lim_{u \to t} \frac{P^{jk}(t, u)}{u - t}. \tag{6.9.8}$$

In the so-called *transition intensity approach* it is assumed that the transition intensities are assigned for all pairs (j, k) such that the direct transition $j \to k$ is possible. From the intensities, via differential equations, probabilities $P^{jk}(t, u)$ can be derived (at least in principle, and in practice numerically) for all states j and k in the state space $\mathscr{S}$.

In the actuarial practice of disability insurance, the intensities should be estimated from statistical data concerning mortality, disability and recovery. Note that for an insurance product providing an annuity in the case of not-necessarily permanent disability the following intensities are required:

$$\mu_t^{ai}, \ \mu_t^{ia}, \ \mu_t^{ad}, \ \mu_t^{id}. \tag{6.9.9}$$

More complicated time-continuous models can be constructed (possibly outside the Markov framework) in order to represent disability duration effects on recovery and mortality. Semi-Markov models can provide the appropriate context. To represent the duration effect, denote by z the time spent in disability in the current disability spell. Then, transition intensities $\mu_{t,z}^{ia}$ and $\mu_{t,z}^{id}$ should be referred to, instead of μ_t^{ia} and μ_t^{id}. This structure has been proposed by the CMI (Continuous Mortality Investigation) Bureau in the UK, to build up a multistate model for covers providing a disability annuity (see Sect. 6.15).

6.10 Practical Actuarial Approaches

The implementation of a rigorous actuarial model for disability insurance requires a lot of statistical data. In the actuarial practice, available data may be scanty (and this in particular happens when new insurance products are concerned). It follows that simplified calculation procedures are often used for pricing and reserving.

Conversely, when statistical data of a given type are available according to a given format, (approximate) calculation procedures are often chosen consistently with type and format. A classification of calculation methods follows, based on the type and format of statistical data supporting pricing and reserving formulae (see Fig. 6.22).

1. *Methods based on the probability of becoming disabled.* The *inception rate* at age y is the frequency with which active individuals become disabled in the year of (exact) age y to $y + 1$. When inception rates are provided by the statistical experience, probabilities of becoming disabled can be estimated. A number of actuarial models based on the probability of becoming disabled, named

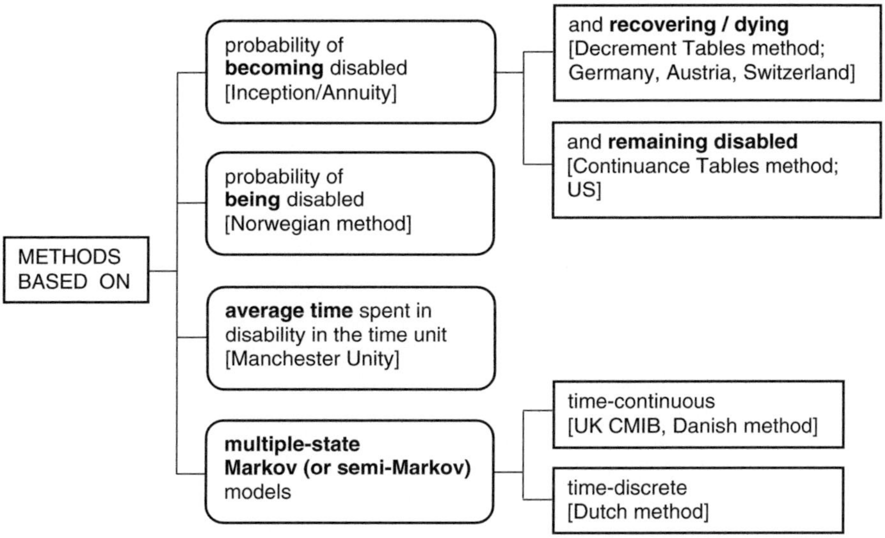

Fig. 6.22 A classification of approaches to actuarial calculations for disability insurance

inception-annuity models, are used in various countries, for example in the US, in Germany, Austria and Switzerland.

a. The method applied in the US market (also known as the *Continuance table method*) is also based on the probabilities of a disabled person remaining in the disability state for a certain length of time, i.e. on a "continuance table".
b. Conversely, the method used in Germany, Austria and Switzerland (also called the *method of decrement tables*) is based on the disabled mortality rates and the recovery rates, i.e. on the rates relating to the two causes of "decrement" from the disability state.

These models are described in Sect. 6.11.

2. *Methods based on the probability of being disabled.* Disability statistical data are often available as *prevalence rates*. The (disability) prevalence rate at age y is defined as the number of disabled individuals aged y (i.e. between exact ages y and $y + 1$) divided by the number of individuals aged y. Prevalence rates lead to the estimation of the probabilities of being disabled at the various ages. Assumptions regarding the policy duration effect on the probability of being disabled are required, as the prevalence rates do not allow for duration effects. An actuarial method based on the probability of being disabled is used, for example, in Norway. See Sect. 6.12.

Remark 1 The inception-annuity method allows for various policy conditions and is more flexible than prevalence-based methods. However, when only prevalence rates are available, the derivation of inception rates from prevalence rates requires assumptions about the mortality of active lives and disabled lives (which are also required for calculating actuarial values of disability annuities) and there could be no experience data to support these assumptions. In these cases, prevalence-based methods can represent a practicable approach. See also Sect. 6.13. ■

3. *Methods based on the expected time spent in disability*. The "disability rate", or "(central) sickness rate", at age y is defined as the average time spent in disability between (exact) ages y and $y + 1$ to the average time lived between ages y and $y + 1$, i.e. the exposure time. This rate can be classified as a *persistency rate*. Sickness rates lead to the estimation of the expected time spent in disability at the various ages. Statistical data arranged in the format of sickness rates underpin the so-called "Manchester Unity" model (or "Friendly Society" model), which has been traditionally used in the UK, until the publication of new statistical data by the Continuous Mortality Investigation (CMI) Bureau in 1991. The Manchester Unity model is described in Sect. 6.14.

Remark 2 The central sickness rate can be compared to the *morbidity coefficient* we have defined in Sect. 5.2.3. Note, however, that in Eq. (5.2.11) it is assumed that the exposure time is equal to 1 for each insured, excluding possible death, lapse or whatever cause of partial exposure. ■

4. *Methods implementing multistate modelling*. Markov (and semi-Markov) multistate models provide a mathematically rigorous and sound framework for analyzing insurances of the person and in particular disability insurance products (see Sect. 6.9). Multistate models can be defined in both a time-continuous and a time-discrete context and offer a powerful tool for interpreting (and criticizing) various practical calculation methods. Moreover, some calculation methods used in actuarial practice directly derive from multistate modelling. Interesting examples in the context of time-continuous implementations are provided by the Danish model and the method proposed by the CMI Bureau in the UK (see Sect. 6.15). In a time-discrete context, the Dutch model shows the possibility, remaining within a Markov framework, of allowing for disability duration effects on recovery and mortality of disabled people (as seen in Sect. 6.8).

Remark 3 The above classification (see Fig. 6.22) is driven by the type of approach adopted in actuarial practice when formulae for premium and reserve calculations were defined, and not by the resulting characteristics of the formulae themselves. An example is given by the inception-annuity methods. These methods can be placed, at least to some extent, in the framework of Markov or semi-Markov models, although no explicit reference to multistate modelling underpinned their structures when these were defined. ■

Some practical calculation methods, belonging to the categories mentioned above, are described in the following sections. For brevity, we only focus on actuarial values leading to single premiums.

6.11 The Inception-Annuity Logic

Formulae in this section can be compared with formulae (6.4.5) and (6.4.7) in Sect. 6.4. Some differences obviously depend on the assumed policy conditions as well as on simplifications and approximations, commonly accepted in the actuarial practice.

Let us start with some formulae adopted in the US. Assume that policy conditions, according to the notation defined in Sect. 6.7, are as follows:

$$\Gamma = \left[0, m, \frac{d}{12}, \infty, \xi - x\right], \tag{6.11.1}$$

where d is the deferred period in months, and ξ denotes some fixed age (e.g. the retirement age). Further, assume that the disability annuity is paid on a monthly basis.

Let $a^{ai}_{x:m|\xi,\frac{d}{12}\rceil}$ denote the actuarial value of the disability benefits (i.e. the single premium according to the equivalence principle) and assume:

$$a^{ai}_{x:m|\xi,\frac{d}{12}\rceil} = \sum_{h=0}^{m-1} v^{h+\frac{1}{2}+\frac{d}{12}}\, {}_h p_x^{aa}\, w_{x+h}\, {}_{\frac{d}{12}} p^{\overline{ii}}_{[x+h+\frac{1}{2}]}\, \ddot{a}^{(12)i}_{[x+h+\frac{1}{2}]+\frac{d}{12}:\xi-x-h-\frac{1}{2}-\frac{d}{12}\rceil}, \tag{6.11.2}$$

where v is the annual discount factor and $\ddot{a}^{(12)i}$ denotes the actuarial value of a disability annuity paid in advance on a monthly basis. Some comments are needed to help understand formula (6.11.2). First, note that $x + h + \frac{1}{2}$ represents the age at disablement (from which the select period starts; see below), while $x + h + \frac{1}{2} + \frac{d}{12}$ represents the age at which the deferment ends and then the annuity commences. The term ${}_h p_x^{aa}$ is the probability for an active person aged x of being active at age $x + h$, whereas w_{x+h} is the probability of becoming disabled between age $x + h$ and $x + h + 1$. Assuming a uniform distribution of the disablement time within the year, the term ${}_{\frac{d}{12}} p^{\overline{ii}}_{[x+h+\frac{1}{2}]}$ represents the probability of remaining disabled from disability inception (at age $x + h + \frac{1}{2}$) to the end of the deferred period. Note that the probability of remaining disabled and hence the actuarial value of the disability annuity are assumed to be inception-select, as indicated by arguments in square brackets.

In order to achieve simpler implementations, formula (6.11.2) can be modified in several ways. For brevity, we only describe three simplifications adopted in the actuarial practice. First, the transition into the disability state is assumed to occur only if the disabled person survives to the end of the deferred period of d months; if death occurs during the deferred period, the transition out of the active state is directly regarded as death. Hence, the survivorship term ${}_{\frac{d}{12}} p^{\overline{ii}}_{[x+h+\frac{1}{2}]}$ is deleted, as the survival is implicitly considered in w_{x+h}. The second simplification is to use ordinary survival probabilities ${}_h p_x$, instead of probabilities of being active ${}_h p_x^{aa}$ (scarcely supported by experience data). This has the effect of increasing the actuarial value of benefits. Finally, the third approximation concerns the factor w_{x+h}. The probability w_y refers to a person aged y. Hence it should be estimated by comparing the number of lives entering into the disability state between age y and $y+1$ to the number of actives (exposed to risk) aged y. In place of w_y, we can use the quantity w'_y which is defined regarding as exposed to risk the average number of actives between age y and $y+1$. The following relation holds:

$$w_y = w'_y \left(1 - \frac{1}{2} q_y^{aa}\right). \tag{6.11.3}$$

By using ordinary survival probabilities, w_y can be approximately expressed as follows:

$$w_y = w'_y \, {}_{\frac{1}{2}}p_y. \tag{6.11.4}$$

The resulting formula for the single premium is then:

$$a^{ai}_{x:m|\xi,\frac{d}{12}\rceil} = \sum_{h=0}^{m-1} v^{h+\frac{1}{2}+\frac{d}{12}} \, {}_{h+\frac{1}{2}}p_x \, w'_{x+h} \, \ddot{a}^{(12)i}_{[x+h+\frac{1}{2}]+\frac{d}{12}:\xi-x-h-\frac{1}{2}-\frac{d}{12}\rceil}. \tag{6.11.5}$$

Now, let us briefly describe some calculation methods used in Europe, based on the inception-annuity model. Assume that the disability annuity is paid on a yearly basis.

A single premium formula used in Germany, for an insurance product with the following policy conditions

$$\Gamma = [0, m, 0, \infty, m], \tag{6.11.6}$$

is given by:

$$a^{ai}_{x:m\rceil} = \sum_{h=0}^{m-1} v^{h+\frac{1}{2}} \, {}_hp_x^{aa} \left(1 - \frac{1}{2} q^{aa}_{x+h}\right) w'_{x+h} \, \frac{1}{2} \left(\ddot{a}^i_{[x+h]:m-h\rceil} + \ddot{a}^i_{[x+h+1]:m-h-1\rceil}\right), \tag{6.11.7}$$

where $\ddot{a}^i_{[x+h]:m-h\rceil}$ denotes the inception-select actuarial value of an annuity payable in advance at policy anniversaries until the insured reactivates or dies, and for m years at most, i.e.:

$$\ddot{a}^i_{[x+h]:m-h\rceil} = \sum_{k=0}^{m-h-1} v^k \, {}_kp^{ii}_{[y]}. \tag{6.11.8}$$

The meaning of the factor w'_{x+h} and the related presence of the factor $\left(1 - \frac{1}{2} q^{aa}_{x+h}\right)$ have been explained above. Note that a linear interpolation formula is used to determine the actuarial value of a disability annuity commencing between ages $x+h$ and $x+h+1$.

A single premium formula used in Austria, assuming a maximum benefit period of s years, that is

$$\Gamma = [0, m, 0, s, \infty] \tag{6.11.9}$$

is as follows:

$$a^{ai}_{x:m;s\rceil} = \sum_{h=0}^{m-1} v^{h+\frac{1}{2}}\, {}_h p^{aa}_x \; w_{x+h} \; \frac{1}{2}\left(\ddot{a}^{i}_{[x+h]:s\rceil} + \ddot{a}^{i}_{[x+h+1]:s\rceil}\right). \tag{6.11.10}$$

Note that the factor $\left(1 - \frac{1}{2}\, q^{aa}_{x+h}\right)$ (see formula (6.11.7)) is omitted since probabilities of the type w_{x+h}, instead of w'_{x+h}, are used.

In Switzerland both total and partial disability are usually covered. In case of partial disability, the disability benefits are graded according to the degree of disablement. Hence, a new factor g_{x+h} ($0 < g_{x+h} \leq 1$), reflecting the average degree of disability among the disabled insureds aged $x+h$, must be introduced in the premium calculation. So, the single premium formula, assuming a maximum benefit period of m years, i.e.

$$\Gamma = [0, m, 0, \infty, m] \tag{6.11.11}$$

is as follows:

$$a^{ai}_{x:m\rceil} = \sum_{h=0}^{m-1} v^{h+\frac{1}{2}}\, {}_{h+\frac{1}{2}} p_x \; w'_{x+h} \; g_{x+h} \; \frac{1}{2}\left(\ddot{a}^{i}_{[x+h]:m-h\rceil} + \ddot{a}^{i}_{[x+h+1]:m-h-1\rceil}\right). \tag{6.11.12}$$

Note the use of an ordinary mortality table, as witnessed by the presence of the probability ${}_{h+\frac{1}{2}} p_x$, instead of ${}_{h+\frac{1}{2}} p^{aa}_x$.

6.12 The Probability of Being Disabled

In a time-discrete context, the actuarial value (at policy issue) of a disability annuity, with a unitary annual amount, payable at policy anniversaries up to the end of the policy term m, is given by:

$$a^{ai}_{x:m\rceil} = \sum_{h=1}^{m} v^h\, {}_h p^{ai}_x, \tag{6.12.1}$$

where ${}_h p^{ai}_x$ denotes the probability that an active insured aged x is disabled at age $x+h$ (see Sect. 6.3). Turning to a time-continuous context, and hence assuming that a continuous annuity is paid at an instantaneous unitary rate while the insured is disabled, the actuarial value is given by:

$$\bar{a}^{ai}_{x:\overline{m}|} = \int_0^m v^t \, {}_t p^{ai}_x \, \mathrm{d}t \,. \tag{6.12.2}$$

Let $j_{(x)+t}$ denote the probability that an individual is disabled at age $x+t$ given that he/she was healthy at age x (i.e. at policy issue) and that he/she is alive at age $x+t$. Note that $j_{(x)+t}$ is a function of the two variables x and t. We have:

$${}_t p^{ai}_x = ({}_t p^{aa}_x + {}_t p^{ai}_x) \, j_{(x)+t}, \tag{6.12.3}$$

and hence

$$\bar{a}^{ai}_{x:\overline{m}|} = \int_0^m v^t \, ({}_t p^{aa}_x + {}_t p^{ai}_x) \, j_{(x)+t} \, \mathrm{d}t \,. \tag{6.12.4}$$

If we approximate the probability for an active individual aged x of being alive at age $x+t$, i.e. ${}_t p^{aa}_x + {}_t p^{ai}_x$, by using a "simple" survival probability ${}_t p_x$ (which does not take into account the state of the insured at age x), we obtain the formula used in the so-called *Norwegian method*:

$$\bar{a}^{ai}_{x:\overline{m}|} = \int_0^m v^t \, {}_t p_x \, j_{(x)+t} \, \mathrm{d}t \,, \tag{6.12.5}$$

which is clearly based on the probability of being disabled.

Remark As regards the availability of data of $j_{(x)+t}$ type, see Sect. 6.13, and Example 6.13.1 in particular. ■

Formula (6.12.5) can also be applied to disability covers allowing for partial disability, in which a graded benefit is paid, depending on the degree of disability. For simplicity, let us assume that the amounts paid are proportional to the degree of disability. Let $g_{x,t}$ denote the expected degree of disability for an individual who is disabled at age $x+t$, being active at age x ($0 < g_{x,t} \leq 1$). The (exact) actuarial value of the disability benefit can easily be obtained, generalizing formula (6.12.2):

$$\bar{a}^{ai}_{x:\overline{m}|} = \int_0^m v^t \, {}_t p^{ai}_x \, g_{x,t} \, \mathrm{d}t \,. \tag{6.12.6}$$

Equating the integrands in (6.12.4) and (6.12.6), we have:

$$j_{(x)+t} = {}_tp_x^{ai} \frac{g_{x,t}}{{}_tp_x^{aa} + {}_tp_x^{ai}}. \tag{6.12.7}$$

Hence, the function $j_{(x)+t}$ can be interpreted as the insured's expected degree of disability at age $x + t$, given that he/she is active at age x and alive at age $x + t$. Then, formula (6.12.5) still holds as an approximation.

6.13 Probability of "Becoming" Versus Probability of "Being"

Assume that disability data are available as *prevalence rates*:

$$\rho_y = \frac{\text{number of people disabled at age } y}{\text{number of people alive at age } y}. \tag{6.13.1}$$

Data of this type can be provided for instance by a social security database, or a public health system database. Assume that the disability prevalence in a given insurance portfolio can (at least approximately) be quantified by referring to the disability prevalence expressed by the database. Nevertheless, the available prevalence rates cannot be directly used for insurance purposes, in particular to assess the probability of being disabled, as they do not assume the individuals were healthy at a given age, viz the age at policy issue.

Example 6.13.1 Consider a portfolio consisting of a cohort entering insurance at age x. In Fig. 6.23 several disability stories of individuals belonging to a reference population are represented. Assume that, at age $x + t$, the reference population consists of 1 000 individuals. Individuals B to H, i.e. 7 individuals, are in the disability state. We than have:

$$\rho_{x+t} = 7/1\,000.$$

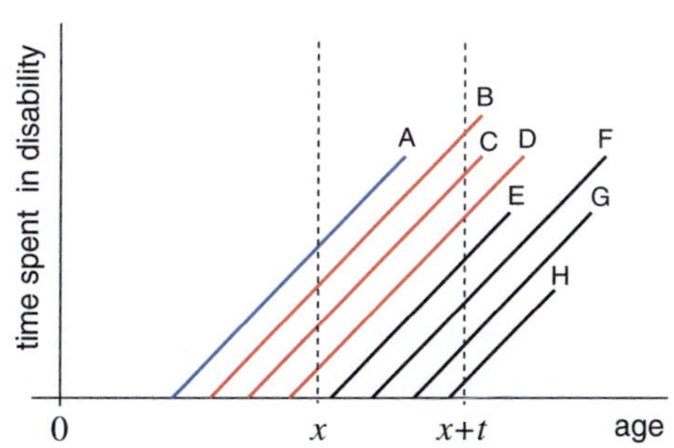

Fig. 6.23 Disability stories

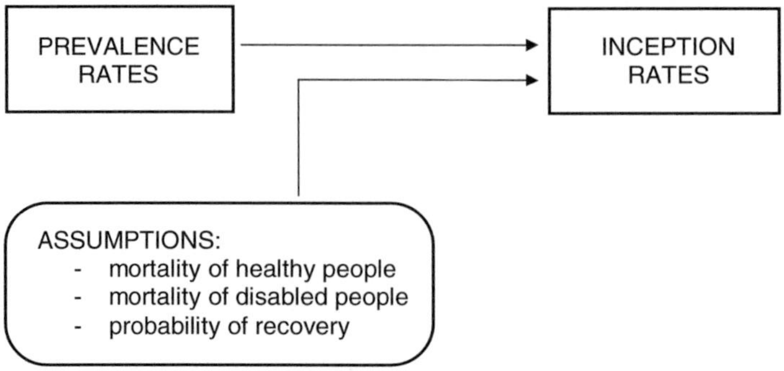

Fig. 6.24 Converting prevalence rates into inception rates

However, individuals B, C and D should not be accounted for when determining the disability prevalence rate at age $x + t$, applicable to the portfolio, because those individuals entered the disability state before age x. Conversely, the disability state of individuals E to H started after age x and is then compatible with the insureds' possible disability stories. Hence, the prevalence rate allowing for the age at entry, that is x, should be 4/1 000 (instead of 7/1 000). □

Two basic approaches are available, in order to obtain, from prevalence data of the type expressed by (6.13.1), data that can be used for calculations in disability insurance. Assume that prevalence rates ρ_{x+t} (possibly smoothed) are available.

1. Define:

$$j_{(x)+t} = \rho_{x+t}\,\alpha(t), \tag{6.13.2}$$

   where $\alpha(t)$ is an adjustment coefficient ($0 < \alpha(t) \leq 1$, increasing as t increases), and assume $j_{(x)+t}$ as the probability of an individual active at age x being disabled at age $x + t$ (see the Norwegian method described in Sect. 6.12).
2. Convert prevalence rates into inception rates, i.e. probabilities of becoming disabled. The point in favor of this approach is the greater flexibility of the inception-annuity method (see Remark 1 in Sect. 6.10); the weak point is given by the set of (critical) assumptions needed to perform the conversion (see Fig. 6.24), in particular related to mortality.

6.14 The Expected Time Spent in Disability

Let us consider formula (6.12.5) and replace $j_{(x)+t}$ with f_{x+t}, which represents the probability that an individual is disabled at age $x + t$ given that he/she is alive at age $x + t$ (disregarding the state at age x, i.e. at policy issue). Let us assume the following approximation for the actuarial value:

$$\bar{a}^{ai}_{x:\overline{m}|} = \int_0^m v^t \, {}_t p_x \, f_{x+t} \, \mathrm{d}t = \frac{1}{\ell_x} \int_0^m v^t \, \ell_{x+t} \, f_{x+t} \, \mathrm{d}t, \tag{6.14.1}$$

where the terms ℓ_y (expected numbers of survivors at age y) are derived from the graduation of an ordinary life table.

It is worth stressing that while in Eq. (6.12.5) the probability $j_{(x)+t}$ is a function of the two variables x and t, the probability f_{x+t} used in (6.14.1) only depends on the attained age $x + t$. It follows that for an insured aged $x + t$ and disabled at that age, any previous age is considered as a possible age of inception of the current disability spell, whilst only disabilities commencing after the entry age x can really lead to the current disability spell (as the insured is assumed to be active at policy issue). A significant overestimate of probabilities f_{x+t} might occur when the relevant estimation is based on claims from a "mature" portfolio. See also Sect. 6.13.

Further, by assuming

$$\ell_{x+h-\frac{1}{2}} \approx \int_0^1 \ell_{x+h-1+\tau} \, \mathrm{d}\tau \tag{6.14.2}$$

we can approximate the actuarial value as follows:

$$\begin{aligned} &\frac{1}{\ell_x} \int_0^m v^t \ell_{x+t} \, f_{x+t} \, \mathrm{d}t \\ &= \frac{1}{\ell_x} \sum_{h=1}^{m} \int_0^1 \ell_{x+h-1+\tau} \, f_{x+h-1+\tau} \, v^{h-1+\tau} \, \mathrm{d}\tau \\ &\approx \frac{1}{\ell_x} \sum_{h=1}^{m} \ell_{x+h-\frac{1}{2}} \, v^{h-\frac{1}{2}} \, \frac{\int_0^1 \ell_{x+h-1+\tau} \, f_{x+h-1+\tau} \, \mathrm{d}\tau}{\int_0^1 \ell_{x+h-1+\tau} \, \mathrm{d}\tau}. \end{aligned} \tag{6.14.3}$$

Let us define the function

$$\theta_y = \frac{\int_0^1 \ell_{y+\tau} \, f_{y+\tau} \, \mathrm{d}\tau}{\int_0^1 \ell_{y+\tau} \, \mathrm{d}\tau}, \tag{6.14.4}$$

which represents the ratio of the expected time spent in disability between ages y and $y + 1$ to the expected time lived between ages y and $y + 1$ (called, in UK actuarial practice, the *central sickness rate*). Using (6.14.4), we obtain the following approximation formula:

$$\bar{a}^{ai}_{x:\overline{m}|} \approx \frac{1}{\ell_x} \sum_{h=1}^{m} \ell_{x+h-\frac{1}{2}} \, v^{h-\frac{1}{2}} \, \theta_{x+h-1}. \tag{6.14.5}$$

Formula (6.14.5) has a central role in the so-called *Manchester Unity model* (or *Friendly Society model*), traditionally used in the UK until the proposal of the CMI Bureau multistate model (see the following section) in 1991. The use of this formula was justified by the availability of experience data in the format of central sickness rates, i.e. ratios of the average time spent in disability to the average time lived in various age intervals. Note that the values $\ell_{x+h-\frac{1}{2}}$ can be obtained by linear interpolation between the items of an ordinary life table.

Formulae useful in disability insurance practice can be implemented through the probability $f_y^{h/k}$ that an insured aged y is disabled, with duration of disability between h and $h+k$. The corresponding sickness rates, $\theta_y^{h/k}$, can be defined as in Eq. (6.14.4). When experience data are available in the format of sickness rates for various h and k, it is possible to define actuarial values of disability benefits with various deferred periods and maximum benefit periods.

6.15 Implementing Multistate Models

Some calculation methods used in actuarial practice implement multistate models for pricing and reserving in relation with disability insurance products. For a brief description of Markov multistate models in the context of disability insurance the reader can refer to Sect. 6.9. Here an implementation of multistate models, including non-Markov models, will be addressed, mainly referring to the model proposed by the CMI Bureau in the UK.

We focus on a multistate model which consists of three states and four transitions, and can then be represented by the graph in Fig. 6.2c. The probabilistic structure is defined by the related transition intensities, with the assumption that:

- the intensities of transition from the active state, i.e. $a \to i$ and $a \to d$, only depend on the attained age;
- the intensities of transition from the disabled state, i.e. the claim termination intensities $i \to a$ and $i \to d$, are assumed to depend on both the attained age and the duration since disability onset, and are hence inception-select intensities.

In any case, no account is taken of other aspects of the past disability story, e.g. the number of previous disability spells. Note that, because of the assumed (and realistic) dependence of the claim termination intensities on the disability duration, the relevant probabilistic model is a semi-Markov model.

Experience data concerning income protection individual policies (and group policies as well) permit the estimation of intensities relating to claim inception and claim termination. The process leading from crude data to actuarial values consists of three steps.

1. First, a *graduation* of observed crude data produces smoothed intensities relating to claim inception and claim termination. Graduated values for the mortality of

active people are taken to be those based on data sets for insured lives with similar characteristics.

2. Secondly, following the *transition intensity approach* transition probabilities are obtained from transition intensities, via numerical integration of simultaneous integro-differential equations.
3. Finally, when transition probabilities are available, actuarial values relating to premiums and reserves can be calculated.

A brief description of steps 2 and 3 follows, and some examples concerning the calculation of transition probabilities from transition intensities and the calculation of actuarial values are discussed.

For the transition intensities, the following notation is used:

$$\mu_{x+t}^{ai},\ \mu_{x+t}^{ad},\ \mu_{x+t,z}^{ia},\ \mu_{x+t,z}^{id}, \tag{6.15.1}$$

where $x + t$ denotes the current age and z the past duration of the current disability spell. For an active insured aged x transition probabilities are denoted as follows:

${}_tp_x^{aa}$ = probability of being active at age $x + t$
${}_tp_x^{ai}$ = probability of being disabled at age $x + t$
${}_tp_x^{ad}$ = probability of being dead at age $x + t$

Further, we define:

${}_{z,t}p_x^{ai}$ = probability of being disabled at age $x + t$ with duration of disability less than or equal to z, $z \le t$ (trivially ${}_{t,t}p_x^{ai} = {}_tp_x^{ai}$).

Let h denote the step size used in numerical procedures for the calculation of probabilities from intensities, for example $h = \frac{1}{100}$. However, it can be useful to link the choice of h to the time unit used to express deferred periods, durations, etc.; if the unit is a week, appropriate choices are, for instance, $h = \frac{1}{104}$ or $h = \frac{1}{156}$. Finally, let us define the following probability:

${}_tp_x^{ai}(r)$ = probability of being disabled at age $x + t$, with duration of disability between $(r - 1)\,h$ and $r\,h$ (with r integer).

The following relations obviously hold:

$${}_tp_x^{ai}(r) = {}_{rh,t}p_x^{ai} - {}_{(r-1)h,t}p_x^{ai}, \tag{6.15.2}$$

$${}_tp_x^{ai} = \sum_{r=1}^{t/h} {}_tp_x^{ai}(r). \tag{6.15.3}$$

Turning to the problem of finding transition probabilities from intensities, first consider, for $r = 1, 2, \ldots$, the following approximate relation:

$$
\begin{aligned}
{}_{t+h}p_x^{ai}(r+1) \approx {}_t p_x^{ai}(r) - \frac{1}{2}h\Big[& {}_t p_x^{ai}(r)\big(\mu^{ia}_{x+t,(r-\frac{1}{2})h} + \mu^{id}_{x+t,(r-\frac{1}{2})h}\big) \\
& + {}_{t+h}p_x^{ai}(r+1)\big(\mu^{ia}_{x+t+h,(r+\frac{1}{2})h} + \mu^{id}_{x+t+h,(r+\frac{1}{2})h}\big)\Big]. \qquad (6.15.4)
\end{aligned}
$$

Formula (6.15.4) can be derived as an approximation formula from a differential equation linking transition probabilities to transition intensities. The derivation is not a trivial matter, so we only present an interpretation of this formula. Moving from probability ${}_t p_x^{ai}(r)$ to ${}_{t+h}p_x^{ai}(r+1)$ means moving the future attained age from $x+t$ to $x+t+h$ and the duration from t to $t+h$. A person who will be disabled at age $x+t$ will still be disabled at age $x+t+h$ if the disability does not terminate, either by recovery or death (assuming that, for h conveniently small, a recovery and a subsequent disablement are impossible). The probability of disability termination is actually given by the second term on the right-hand side of (6.15.4).

Assume now that x is given, e.g. the age at policy issue. Provided that we can calculate ${}_t p_x^{ai}(r)$ for $r = 1$ and for each t, we can recursively calculate ${}_t p_x^{ai}(r)$ for each t and for $r = 2, 3, \ldots$. From a computational point of view, it is convenient to assume that the claim termination intensities only depend on the age $x+t$ for $r \geq r'$; the value r' can be suggested by statistical evidence.

As a second example, consider the following approximate recurrence relation:

$$
\begin{aligned}
{}_{t+h}p_x^{aa} \approx {}_t p_x^{aa} & - \frac{1}{2}h\Big[{}_t p_x^{aa}\big(\mu^{ai}_{x+t} + \mu^{ad}_{x+t}\big) + {}_{t+h}p_x^{aa}\big(\mu^{ai}_{x+t+h} + \mu^{ad}_{x+t+h}\big)\Big] \\
& + \frac{1}{2}h\sum_r\Big[{}_t p_x^{ai}(r)\,\mu^{ia}_{x+t,(r-\frac{1}{2})h} + {}_{t+h}p_x^{ai}(r)\,\mu^{ia}_{x+t+h,(r-\frac{1}{2})h}\Big]. \qquad (6.15.5)
\end{aligned}
$$

The interpretation of (6.15.5) is as follows. A person will be active at age $x+t$ if:

- he/she is active at age x and he/she does not leave this state because of disablement or death, the probability of this event being represented by the second term on the right-hand side of (6.15.5);
- he/she is disabled at age x and he/she recovers, the probability of this event being represented by the third term on the right-hand side of (6.15.5), which allows for the duration of the disability spell.

We now turn to some examples concerning the calculation of actuarial values of continuous annuities. Continuous annuities closely approximate to annuities payable in strict proportion to the duration of disability. In all the cases, we refer to an insured who is active at age x (e.g. at policy issue), and we assume that the annuity is payable while the insured is disabled, up to the policy term m.

Let us consider an annuity payable at an instantaneous rate of 1 per annum. The relevant actuarial value, $\bar{a}^{ai}_{x:\overline{m}|}$, is given by formula (6.12.2) and can be approximately calculated using, for example, the trapezium rule. With a step h, we have:

$$
\bar{a}^{ai}_{x:\overline{m}|} \approx \sum_{j=0}^{\frac{m}{h}-1} \frac{1}{2}h\big[v^{jh}\,{}_{jh}p_x^{ai} + v^{(j+1)h}\,{}_{(j+1)h}p_x^{ai}\big]. \qquad (6.15.6)
$$

Formula (6.15.6) can be generalized in order to allow for policy conditions including a deferred period, or a benefit amount varying as the duration of disability increases. We do not address this feature.

The actuarial value of a continuous annuity payable at an instantaneous unitary rate while the insured is active and for m' years at most is given by:

$$\bar{a}^{aa}_{x:\overline{m'}|} = \int_{0}^{m'} v^t \, {}_t p^{aa}_x \, \mathrm{d}t. \tag{6.15.7}$$

Formula (6.15.7) can be adopted for calculating periodic level premiums, which are waived during disability spells. The relevant approximation based on the trapezium rule is as follows:

$$\bar{a}^{aa}_{x:\overline{m'}|} \approx \sum_{j=0}^{\frac{m'}{h}-1} \frac{1}{2} h \left[v^{jh} \, {}_{jh} p^{aa}_x + v^{(j+1)h} \, {}_{(j+1)h} p^{aa}_x \right]. \tag{6.15.8}$$

6.16 Actuarial Models for LTCI: An Introduction

In this section we focus on long-term care insurance (LTCI) products which provide graded annuity benefits, i.e. benefits whose amount is graded according to the insured's disability degree. In particular, we only address stand-alone covers (see Sects. 3.6.3).

6.16.1 A Basic Biometric Model

When the disability degree is expressed in terms of a (small) number of disability states (say, 2 to 4; see Sect. 3.6.1), the actuarial model can be based on a multistate structure, similar to those addressed in Sect. 6.3.

We note that:

- more than one disability state must be considered to represent graded benefits (for example, see Fig. 6.25a in which two disability states are assumed);
- conversely, a simplified structure can be adopted if the possibility of recovery is disregarded, consistently with its very low probability (see Fig. 6.25b).

Following the approach adopted in Sect. 6.3 (see in particular Sect. 6.3.2), we start defining one-year transition probabilities. In particular, we refer to probabilities given by the multistate model represented in Fig. 6.25b, where i' and i'' denote the states corresponding to the lower and the higher severity of disability, respectively.

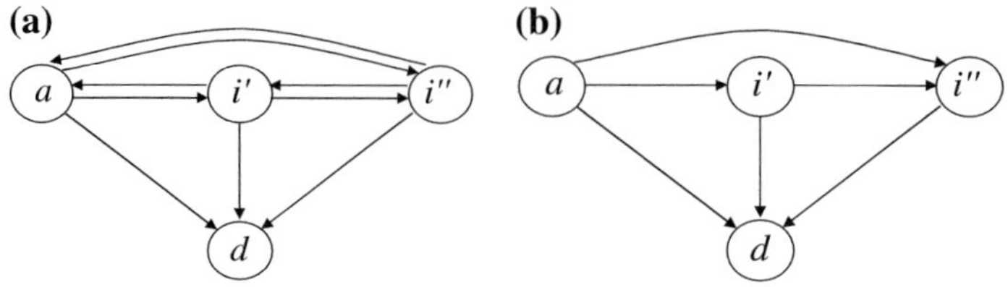

Fig. 6.25 Four-state models for LTC

The one-year transition probabilities constitute the stochastic matrix represented by Table 6.7.

From the one-year transition probabilities, multi-year transition probabilities can be derived by adopting the procedure described in Sect. 6.3.3 (and the Chapman–Kolmogorov equations in particular):

$$ {}_h p_y^{aa} = {}_{h-1}p_y^{aa}\; p_{y+h-1}^{aa}, \tag{6.16.1a}$$

$$ {}_h p_y^{ai'} = {}_{h-1}p_y^{ai'}\; p_{y+h-1}^{i'i'} + {}_{h-1}p_y^{aa}\; p_{y+h-1}^{ai'}, \tag{6.16.1b}$$

$$ {}_h p_y^{ai''} = {}_{h-1}p_y^{ai''}\; p_{y+h-1}^{i''i''} + {}_{h-1}p_y^{ai'}\; p_{y+h-1}^{i'i''} + {}_{h-1}p_y^{aa}\; p_{y+h-1}^{ai''}. \tag{6.16.1c}$$

The following relations also hold:

$$ {}_h p_y^{ai'} = \sum_{r=1}^{h} {}_{h-r}p_y^{aa}\; p_{y+h-r}^{ai'} \prod_{g=1}^{r-1} p_{y+h-r+g}^{i'i'}, \tag{6.16.2a}$$

$$ {}_h p_y^{ai''} = \sum_{r=1}^{h} \Big[{}_{h-r}p_y^{aa}\; p_{y+h-r}^{ai''} \prod_{g=1}^{r-1} p_{y+h-r+g}^{i''i''} + {}_{h-r}p_y^{ai'}\; p_{y+h-r}^{i'i''} \prod_{g=1}^{r-1} p_{y+h-r+g}^{i''i''} \Big]. \tag{6.16.2b}$$

We note that (6.16.2a) replicates Eq. (6.3.23), whereas (6.16.2b) generalizes the same equation by considering two disability states instead of one.

Table 6.7 Conditional probabilities related to LTCI products

State at age y	State at age $y+1$			
	a	i'	i''	d
a	p_y^{aa}	$p_y^{ai'}$	$p_y^{ai''}$	q_y^{a}
i'	0	$p_y^{i'i'}$	$p_y^{i'i''}$	$q_y^{i'}$
i''	0	0	$p_y^{i''i''}$	$q_y^{i''}$
d	0	0	0	1

6.16.2 Actuarial Values and Premiums

Assume the following annual benefits:

$b' = 1$, if the insured is in the disability state i';
$b'' = 1 + \beta$, if the insured is in the disability state i'' $(\beta > 0)$.

To help interpret the following formulae (6.16.3) and (6.16.4), we note what follows.

- The state i' can only be entered from state a, that is, with the transition $a \to i'$.
- The state i'' can be entered from the state a directly, that is, with the transition $a \to i''$, or from state a via state i', that is, with the two transitions $a \to i'$ and $i' \to i''$; in the latter case, two years are required, because of the assumption of one transition at most in a year.

See also Table 6.7.

Let $a_x^{ai'i''}(\beta)$ denote the actuarial value (and hence, according to the equivalence principle, the single premium) of the LTCI product sold to an active individual age x at policy issue, providing the benefits 1 and $1+\beta$ accordingly to the disability state. Benefits are assumed payable at policy anniversaries. We have:

$$\begin{aligned}
a_x^{ai'i''}(\beta) &= \sum_{h=1}^{+\infty} \left({}_hp_x^{ai'} + (1+\beta)\, {}_hp_x^{ai''} \right) v^h \\
&= \sum_{h=1}^{+\infty} \Bigg[\sum_{r=1}^{h} {}_{h-r}p_x^{aa}\, p_{x+h-r}^{ai'} \prod_{g=1}^{r-1} p_{x+h-r+g}^{i'i'} \\
&\quad + (1+\beta) \sum_{r=1}^{h} {}_{h-r}p_x^{aa}\, p_{x+h-r}^{ai''} \prod_{g=1}^{r-1} p_{x+h-r+g}^{i''i''} \\
&\quad + (1+\beta) \sum_{r=1}^{h-1} {}_{h-r}p_x^{ai'}\, p_{x+h-r}^{i'i''} \prod_{g=1}^{r-1} p_{x+h-r+g}^{i''i''} \Bigg] v^h .
\end{aligned} \tag{6.16.3}$$

After some manipulations, and with obvious definitions for $\ddot{a}_{x+j}^{i'}$ and $\ddot{a}_{x+j}^{i''}$ (see Eq. (6.4.5) by analogy), we finally obtain:

$$\begin{aligned}
a_x^{ai'i''}(\beta) &= \sum_{j=1}^{+\infty} {}_{j-1}p_x^{aa}\, p_{x+j-1}^{ai'}\, v^j\, \ddot{a}_{x+j}^{i'} \\
&\quad + (1+\beta) \sum_{j=1}^{+\infty} {}_{j-1}p_x^{aa}\, p_{x+j-1}^{ai''}\, v^j\, \ddot{a}_{x+j}^{i''} \\
&\quad + (1+\beta) \sum_{j=2}^{+\infty} {}_{j-1}p_x^{ai'}\, p_{x+j-1}^{i'i''}\, v^j\, \ddot{a}_{x+j}^{i''} .
\end{aligned} \tag{6.16.4}$$

We note that (6.16.4) is an *inception-annuity* formula, which generalizes (6.4.5) by considering two disability states.

Assume that annual level premiums are paid for m' years, while the insured is healthy (i.e. in state a). Then, the annual level premium P is given by:

$$P = \frac{a_x^{ai'i''}(\beta)}{\ddot{a}_{x:\overline{m'}|}^{aa}}, \tag{6.16.5}$$

where $\ddot{a}_{x:\overline{m'}|}^{aa}$ is defined as in (6.4.13).

6.16.3 Longevity Risk Issues

It is well known that in all lifelong living benefits (i.e. life annuities, lifelong sickness insurance covers, etc.) the insurer bears the longevity risk, and in particular its systematic component (the *aggregate longevity risk*) caused by the possibility that all the insureds live, on average, longer than expected (see also Sect. 2.3). This risk component is undiversifiable via pooling, i.e. inside the traditional insurance-reinsurance process.

In the case of health insurance products, such as LTC covers, risk emerges further from uncertainty concerning the time spent in the disability state. Actually, when living benefits are paid in the case of disability (senescent disability in particular) it is not only important how long one lives, but also how long he/she lives in a condition of disability.

Although it is reasonable to assume a relationship between mortality and morbidity or disability, the relevant definition is difficult due to the complexity of such a link and the impossibility of defining and measuring disability objectively. Three main theories have been proposed about the evolution of senescent disability (see references in Sect. 6.18). The most important features of the three theories can be expressed in terms of the evolution of the total life expectancy (TLE) and the disability-free life expectancy (DFLE) (both expectancies can be considered at birth or at some given adult age), as represented in Fig. 6.26. The expected length of the (senescent) disability period is then given by TLE-DFLE.

Ideas underlying the three theories are as follows.

- *Compression theory*: chronic degenerative diseases will be postponed until the latest years of life thanks to medical advances. Assuming there is a maximum limit for the total life expectancy, these improvements will result in a compression of the period of disability (see Fig. 6.26a).
- *Equilibrium theory*: most of the changes in mortality are related to specific pathologies. The onset of chronic degenerative diseases and disability will be postponed and the time of death as well, resulting in a more or less constant spread between TLE and DFLE (see Fig. 6.26b).

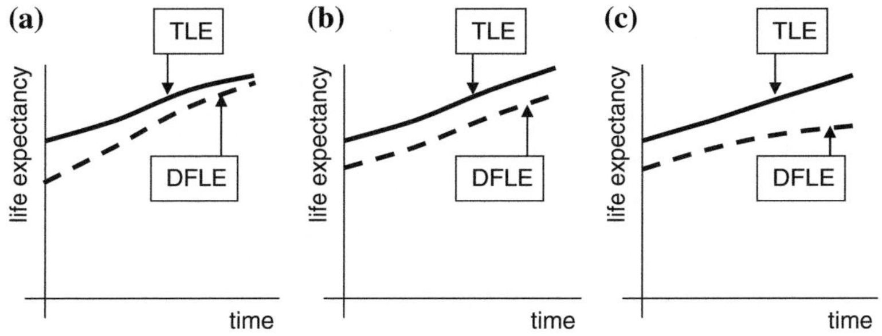

Fig. 6.26 Trends in total life expectancy (TLE) and disability-free life expectancy (DFLE), according to different theories. (**a**) Compression theory. (**b**) Equilibrium theory. (**c**) Pandemic theory

- *Pandemic theory*: improvements in mortality are not accompanied by a decrease in disability rates, and this results in an increasing spread between TLE and DFLE (see Fig. 6.26c). Hence, the number of disabled people will increase steadily.

In order to assess the risks inherent in LTC covers, uncertainty in future mortality and disability trends should explicitly be taken into account. To this end, several scenarios must be considered, each one including a specific projection of mortality and disability trends, which represents a possible realization of the actual future scenario.

It is beyond the scope of this book to deal with these complex problems which, however, should be carefully considered in the risk identification, risk assessment and impact assessment steps in the risk management process. The interested reader can refer, in particular, to some papers and reports cited in Sect. 6.18.

6.17 Mortality of Disabled People

Some aspects of mortality among disabled people and the related formal representation are addressed in this section.

6.17.1 Mortality Data: Some Critical Aspects

Statistical observations show that both frequencies of recovery and death of disabled people depend on the past duration of disability (see Sect. 6.8). Hence, probabilities of death adopted in actuarial models should rely on inception-select mortality data, that is, taking into account the time spent in disability, rather than "aggregate" mortality data (with respect to time spent in disability).

Of course mortality also depends on the cause (e.g. accident versus sickness) and severity (partial versus total, temporary versus permanent) of disability. It is worth stressing that in different countries the eligibility to disability benefits (both in private insurance and in social security as well) can be different, in particular according to legislation, usual policy conditions (see also Sect. 3.5), market practice, and so on. Reasonable groupings can be defined according to the specific insurance covers addressed.

Finally, it should be recalled (see Sect. 6.5) that disability benefits (for example in IP and LTC insurance products) are living benefits, that is, benefits are payable as long as the insured is alive (and disabled as well). Thus, a safe-side assessment of the insurer's liabilities related to disability and LTC annuities requires that the mortality of disabled people should not be overestimated.

6.17.2 Modelling Extra-Mortality

Disabled people constitute, as regards mortality, *substandard risks*. Hence, mortality of disabled people contains an "extra-mortality" component, and can be represented either as a specific mortality (via appropriate numerical tables or parametric mortality laws) or via adjustments to the standard age-pattern of mortality. We now focus on these alternatives.

Mortality of disabled people, in terms of one-year probability of death, has been denoted in Sect. 6.3.2 by the symbol q_y^i (disregarding the inception-selection effect). We now adopt the notation $q_{y,z}^{(k)}$, in order to explicitly refer also to the disability cause or severity, and the time since disability inception as well.

If a mortality law Ψ has been chosen, then we have:

$$q_{y,z}^{(k)} = \Psi(y, z; k), \tag{6.17.1}$$

where:

- y is the current age;
- z is the time elapsed since disability inception, that is, the past duration of the current disability spell;
- k denotes a category of disability, expressing in particular the severity of the disability, and entering the function Ψ via appropriate parameters.

Note that, according to (6.17.1):

- disabled mortality is described via inception-select probabilities, and hence it depends on the past duration of disability (see also Sect. 6.8);
- even if causes of disability are not explicitly allowed for, the severity of disability and the related impact on mortality can be expressed via parameter k.

Instead of relying on the use of specific functions $\Psi(y, z; k)$, the mortality of disabled people can be expressed in relation to the average (or standard) mortality (in a population, or in a portfolio, or pension plan). This allows us to deal only with one life table (or one mortality law), properly adjusted when applied to disabled people.

Thus, if q_y denotes the annual probability of death for an individual age y according to the age-pattern of mortality taken as the standard, the "adjusted" probability of death, $q_{y,z}^{(k)}$, is assumed to be expressed as follows:

$$q_{y,z}^{(k)} = \Phi(q_y, z; k), \tag{6.17.2}$$

where time z since disability inception and disability category k will enter the function Φ via appropriate parameters.

A (rather) general model of type (6.17.2) is the following one:

$$\Phi(q_y, z; k) = A_z^{(k)} \, q_{y+B^{(k)}} + C_z^{(k)}. \tag{6.17.3}$$

Note that:

- the parameters A, B and C are category-dependent;
- in general, A and C are functions of time z, and hence express the duration effect on mortality;
- the parameter B is a "years to age" addition, also called the "age-shift" parameter.

Several models adopted in actuarial practice constitute particular implementations of model (6.17.3).

If we assume $B^{(k)} = 0$ for any k, and $A_z^{(k)} = \bar{A}^{(k)}$, $C_z^{(k)} = \bar{C}^{(k)}$ for all z, we obtain the *linear model* (with flat parameters):

$$\Phi^{[L]}(q_y; k) = \bar{A}^{(k)} \, q_y + \bar{C}^{(k)}. \tag{6.17.4}$$

Note that, according to (6.17.4), the duration effect is disregarded and hence aggregate probabilities are adopted to express the age-pattern of mortality of disabled people.

In particular, by setting $\bar{A}^{(k)} = 1$ (and $\bar{C}^{(k)} > 0$) for any k we find the *additive model*:

$$\Phi^{[A]}(q_y; k) = q_y + \bar{C}^{(k)}, \tag{6.17.5}$$

which expresses a constant extra-mortality.

Conversely, with $\bar{C}^{(k)} = 0$ (and $\bar{A}^{(k)} > 1$) for any k, we have the *multiplicative model*:

$$\Phi^{[M]}(q_y; k) = \bar{A}^{(k)} \, q_y, \tag{6.17.6}$$

which expresses an increasing extra-mortality (given that the probability of death q_y increases as the age increases).

Finally, assuming $A_z^{(k)} = 1$ and $C_z^{(k)} = 0$ for all z and any k, Eq. (6.17.3) results in the *age-shift model*:

$$\Phi^{[S]}(q_y; k) = q_{y+B^{(k)}}. \tag{6.17.7}$$

If probabilities q_y approximately follow an exponential pattern, this model can be considered as an approximation to the multiplicative model (6.17.6).

Models (6.17.5), (6.17.6) and (6.17.7) are also frequently used in life insurance for pricing death benefits when the insured is classified as a substandard risk.

6.18 Suggestions for Further Reading

A number of calculation methods for disability insurance covers (and related products, e.g. Critical Illness insurance and long-term care insurance) are discussed in Haberman and Pitacco (1999); in this textbook the practical calculation methods adopted in various countries for pricing and reserving are presented and discussed within the context of multistate modelling; a comprehensive list of references, mainly of interest to actuaries, is provided.

The actuarial textbooks by Bowers et al. (1997) and by Dickson et al. (2013) also address pricing and reserving for disability insurance, in the broader framework of life contingencies.

Disability covers commonly sold in the US are described in Bartleson (1968), O'Grady (1988), Black and Skipper (2000) (which also describes life insurance), and Bluhm (1992) (which focusses on group insurance).

European disability products and the relevant actuarial methods are described in many papers which have appeared in actuarial journals. Disability insurance in the UK is described, for example, in Mackay (1993) and Sanders and Silby (1988). Disability covers sold in the Netherlands are illustrated in Gregorius (1993). For information on disability insurance in Germany, Austria and Switzerland, readers should consult Segerer (1993).

References regarding models adopted in disability actuarial practice can be split into two categories. In the first category we collect contributions mainly related to "local" calculation methods. The inception-annuity method adopted in the US is described in many textbooks; in particular readers are referred to Bowers et al. (1997). For information on European implementations of the inception-annuity method, readers should consult Segerer (1993). The Danish model is briefly described in Ramlau-Hansen (1991). The Norwegian method (the so-called j-method) is described in particular in Sand and Riis (1980). The Dutch model is discussed in Gregorius (1993).

The CMI Bureau model, relating to actuarial practice in the UK, is described in CMI (1991); the reader should also consult Hertzman (1993) and Waters (1989). The traditional British "Manchester Unity" model is described in several actuarial textbooks; see for example Benjamin and Pollard (1993). A probabilistic critique of the quantities involved in actuarial calculations according to this method is presented by Haberman (1988).

Interesting disability data referred to the UK market, as well as various methodological issues, can be found on CMI Working papers and Reports, available on: http://www.actuaries.org.uk/research-and-resources/pages/continuous-mortality-investigation-working-papers and http://www.actuaries.org.uk/research-and-resources/pages/continuous-mortality-investigation-reports respectively.

Turning to "country independent" studies, we can define a second category including papers and books dealing with general aspects of actuarial models for disability insurance. The reader interested in comparing different calculation techniques for insurance policies providing disability annuities should consult Hamilton-Jones (1972) and Mattsson (1977). A concise presentation of products providing disability annuities and the relevant calculation methods is provided by Pitacco (2004a, b) respectively. IBNR reserves in disability insurance are dealt with in Waters (1992).

Actuarial models, and in particular approximation methods, for evaluating disability lump sums are dealt with in Bull (1980).

Several calculations methods are presented and discussed in the context of multistate modelling in Pitacco (1995). Multistate models are specifically adopted in Aro et al. (2013) and Christiansen (2012).

The "disability process", also interesting sickness insurance and accident insurance, is addressed in Olivieri and Pitacco (2009) within the framework of multistate modelling. As regards sociomedical aspects of disablement, see Verbrugge and Jette (1994) and the references therein.

A stochastic approach to the evaluation of insurance products providing disability annuities is proposed in Haberman et al. (2004), where the impact of both biometric and financial risks is accounted for.

Literature devoted to actuarial issues in LTCI includes: American Academy of Actuaries (1999), Dullaway and Elliott (1998), Gatenby (1991), Jones (2004), Leung (2004), Pitacco (1994, 1999), and Society of Actuaries Long-Term Care Insurance Valuation Methods Task Force (1995).

Various aspects of LTC services provided by Continuing Care Retirement Communities (including methodological and statistical issues) are analyzed by Jones (1995, 1996, 1997a, b).

Assessment of disability severity (via ADL and IADL methods in particular) is analyzed in Lawton and Brody (1969), Martin and Elliot (1992), and McDowell (2006). A multistate model approach to functional disability is adopted by Fong et al. (2013).

The reader interested in the evolution of the total and the disability-free life expectancy can refer to Browne (2011), and the original contributions by Fries (1980), Gruenberg (1977), Kramer (1980) and Manton (1982). The impact of uncertainty in future longevity and healthy longevity on LTCI portfolio results is analyzed and quantified in Olivieri and Ferri (2003) and Olivieri and Pitacco (2002a).

Mortality of disabled people is addressed in Pitacco (2012). A comprehensive model representing the mortality of extra-risks has been proposed by Ainslie (2000), while focussing on life annuities for impaired lives. Ellingsen (2010) addresses mortality among disabled pensioners, while Sanchez-Delgado et al. (2009) focus on mortality of disabled people in the Spanish population. Mortality of disabled people is also considered by Rickayzen (2007) in the context of disability-linked annuities, and Rickayzen and Walsh (2002) in the framework of long-term care need analysis.

To conclude this bibliographic review, we cite some important contributions in the field of Markov (and semi-Markov) models and their applications to the mathematics of insurances of the person. The basic mathematics of what we now call a Markov model was developed during the 18th century (see Seal (1977)). A systematic approach to disability actuarial problems, consistent with the Markov assumption, dates back to the beginning of the 1900s (see Haberman (1996)), and, finally, an explicit use of the mathematics of Markov multistate models dates back to the second half of the 20th century. The first studies concerning the Markov approach to actuarial problems in life and disability insurance seem to be due to Franckx (1963), Daboni (1964) and Amsler (1968).

The seminal paper by Hoem (1969) places life and other contingencies within the framework of a general, unified, probabilistic theory relying on the Markov assumption; a time-continuous approach is adopted, and formulae and theorems for actuarial values, premiums and reserves are derived. More recent contributions to life insurance and related fields, based on multistate models, are given by Amsler (1988), Hoem (1988) and Waters (1984).

The first application of semi-Markov models to disability insurance seems to be due to Janssen (1966). The use of semi-Markov processes in actuarial science and demography is discussed by Hoem (1972).

References

Although the book is "teaching" rather than "research oriented", many scientific and professional references are listed, such as papers in scientific journals and conference proceedings, working papers, technical reports, etc. This material can provide substantial help especially if, for specific topics, textbooks are not available, or not updated.

Where links are provided, they were active as of the time this book was completed but may have been updated since then.

Ainslie, R. (2000). Annuity and insurance products for impaired lives. The Staple Inn Actuarial Society, London. Available at: http://www.sias.org.uk/siaspapers/search/view_paper?id=ImpairedLives

Alsina, C., Trillas, E., and Moraga, C. (2003). Combining degrees of impairment: The case of the index of Balthazard. Mathware & Soft Computing, 10:23–41.

American Academy of Actuaries (1999). Long-term care. Actuarial issues in designing voluntary federal-private LTC insurance programs. Public Policy Monograph, Washington. Available at: http://www.actuary.org/pdf/health/LTC.pdf

Amsler, M. H. (1968). Le chaînes de Markov des assurances vie, invalidité et maladie. In Transactions of the 18th International Congress of Actuaries, volume 5, pages 731–746, München.

Amsler, M. H. (1988). Sur la modélisation de risques vie par les chaînes de Markov. In Transactions of the 23rd International Congress of Actuaries, volume 3, pages 1–17, Helsinki.

Aro, H., Djehiche, B., Löfdahl, B. (2013). Stochastic modelling of disability insurance in a multi-period framework. Scandinavian Actuarial Journal. Available at: http://dx.doi.org/10.1080/03461238.2013.779594

Bartleson, E. L. (1968). Health Insurance. The Society of Actuaries, Illinois, USA.

Baumann, F., Meier, V., Werding, M. (2004). Transferable ageing provisions in individual health insurance contracts. CESIFO Working Paper No. 1116 - Category 9: Industrial Organization. Available at: http://www.cesifo-group.de/ifoHome/publications/working-papers/CESifoWP.html

Benjamin, B. and Pollard, J. H. (1993). The Analysis of Mortality and Other Actuarial Statistics. The Institute of Actuaries, Oxford.

E. Pitacco, *Health Insurance*, EAA Series,
DOI 10.1007/978-3-319-12235-9

Bernet, P. M. and Getzen, T. E. (2004). Health insurance. In Teugels, J. L. and Sundt, B., editors, Encyclopedia of Actuarial Science, volume 2, pages 799–809. John Wiley & Sons.

Black, K. and Skipper, H. D. (2000). Life & Health Insurance. Prentice Hall, New Jersey.

Bluhm, W. F., editor (1992). Group Insurance. ACTEX Publications, Winsted, Connecticut, USA.

Bowers, N. L., Gerber, H. U., Hickman, J. C., Jones, D. A., and Nesbitt, C. J. (1997). Actuarial Mathematics. The Society of Actuaries, Schaumburg, Illinois, USA.

Brecht, S. B., Fein, S., and Hollinger-Smith, L. (2009). Preparing for the future: trends in Continuing Care Retirement Communities. Seniors Housing & Care Journal, 16(1):47–62.

Brink, A. (2010). Practical example of a split benefit accelerated Critical Illness insurance product. In *Transactions of the 29th International Congress of Actuaries*, Cape Town, South Africa. Available at: http://www.actuaries.org/EVENTS/Congresses/Cape_Town/Papers/Health%20(IAAHS)/33_final%20paper_Brink.pdf.

Brown, J. and Warshawsky, M. (2013). The Life Care annuity: a new empirical examination of an insurance innovation that addresses problems in the markets for life annuity and Long-Term Care insurance. The Journal of Risk and Insurance, 8(3):677–703.

Browne, B. (2011). Healthy longevity. Working Paper, Mortality Working Group of the IAA. Available at:http://www.actuaries.org/CTTEES_TFM/Documents/Healthy%20Longevity%20paper%20Bridget%20Browne.pdf

Bull, O. (1980). Premium calculations for two disability lump sum contracts. In Transactions of the 21st International Congress of Actuaries, volume 3, pages 45–51, Zürich-Lausanne.

Chen, Y. (1994). Long-term care: an intragenerational social insurance model. The Geneva Papers on Risk and Insurance, 19(73):490–495.

Christiansen, M. C. (2012). Multistate models in health insurance. AStA Advances in Statistical Analysis, 96:155–186.

Christianson, J. B., Parente, S. T., and Taylor, R. (2002). Defined-contribution health insurance products: development and prospects. Health Affairs, 21(1):49–64. Available at: http://content.healthaffairs.org/content/21/1/49.full.pdf+html

Cichon, M., Newbrander, W., Yamabana, H., Weber, A., Normand, C., Dror, D., Preker, A. (1999). Modelling in health care finance: A compendium of quantitative techniques for health care financing. International Labour Office, Geneva. Available at: http://www.ilo.org/secsoc/information-resources/publications-and-tools/quantitative-methods/WCMS_209903/lang-en/index.htm

CMI (1991). The analysis of Permanent Health Insurance data. CMI Report 12, Continuous Mortality Investigation, Institute of Actuaries and Faculty of Actuaries. Available at: http://www.actuaries.org.uk/research-and-resources/documents/cmi-report-12-whole-volume

CMI (2006). Analysis of individual Income Protection experience by cause of disability. CMI Working Paper 23, Continuous Mortality Investigation, Institute of Actuaries and Faculty of Actuaries. Available at: http://www.actuaries.org.uk/research-and-resources/pages/cmi-working-paper-23

Colombo, F. and Tapay, N. (2004). Private health insurance in OECD countries: The benefits and costs for individuals and health systems. OECD Health Working Papers, No. 15, OECD Publishing. Available at: http://www.oecd.org/els/health-systems/33698043.pdf

Costa-Font, J. and Courbage, C., editors (2012). Financing Long-Term Care in Europe. Institutions, Markets and Models. Palgrave Macmillan.

Cucuta, C. (2002). Defined contribution health plans: Ready or not, here they come. Marquette Elder's Advisor, 4(1):32–41. Available at: http://scholarship.law.marquette.edu/elders/vol4/iss1/6

Daboni, L. (1964). Modelli di rendite aleatorie. Giornale dell'Istituto Italiano degli Attuari, 27:273–296.

Dash, A. and Grimshaw, D. (1993). Dread Disease cover. An actuarial perspective. Journal of the Staple Inn Actuarial Society, 33:149–193. Available at: http://www.actuaries.org.uk/research-and-resources/documents/dread-disease-cover-actuarial-perspective

de Braaf, K. (2013). Multiple Critical Illness benefits from market needs to product solutions. Asia-Pacific experience. SCOR Global Life Newsletter. Available at: http://www.scor.com/en/sgrc/life/disease-health-disability/item/2550.html?lout=sgrc

Dickson, D. C. M., Hardy, M. R., and Waters, H. R. (2013). Actuarial Mathematics for Life Contingent Risks. Cambridge University Press, 2nd edition.

Dror, D. M. and Jacquier, C. (1999). Micro-insurance: Extending health insurance to the excluded. International Social Security Review, 52(1):71–97.

Dullaway, D. and Elliott, S. (1998). Long-term care insurance: a guide to product design and pricing. The Staple Inn Actuarial Society, London. Available at: http://www.actuaries.org.uk/research-and-resources/documents/long-term-care-insurance-guide-product-design-and-pricing

Ellingsen, T. M. (2010). Mortality among disability pensioners. In Transactions of the 29th International Congress of Actuaries, Cape Town, South Africa. Available at: http://www.actuaries.org/EVENTS/Congresses/Cape_Town/Papers/Health%20(IAAHS)/03_final%20paper_Ellingsen.pdf

Elliott, S., Golds, S., Sissons, I., Wilson, J. K. (2014). Long Term Care - A review of global funding models. (A working paper). British Actuarial Journal, 19:55–115.

Fabrizio, E. and Gratton, W. K. (1994). Pricing Dread Disease insurance. Institute of Actuaries of Australia, Sydney.

Fong, J. H., Shao, A. W., and Sherris, M. (2013). Multi-state actuarial models of functional disability. CEPAR Working Paper 2013/16. Available at: http://www.cepar.edu.au/working-papers/working-papers-2013.aspx

Franckx, E. (1963). Essai d'une théorie opérationelle des risques markoviens. Quaderni dell'Istituto di Matematica Finanziaria dell'Università di Trieste, No. 11.

Fries, J. F. (1980). Aging, natural death, and the compression of morbidity. The New England Journal of Medicine, 303(3):130–135.

Fürhaupter, R. and Brechtmann, C. (2002). A survey of the private health insurance in Germany. In Transactions of the 27th International Congress of Actuaries, Cancun, Mexico. Available at: http://www.actuaries.org/EVENTS/Congresses/Cancun/health_subject/health_28_furhaupter.pdf

Gatenby, P. (1991). Long Term Care. The Staple Inn Actuarial Society, London. Available at: http://www.sias.org.uk/siaspapers/search/view_paper?id=ltc.pdf

Gleckman, H. (2010). Long Term Care financing reform: Lessons from the U.S. and abroad. The Commonwealth Fund. Available at: http://www.commonwealthfund.org/~/media/Files/Publications/Fund%20Report/2010/Feb/1368_Gleckman_longterm_care_financing_reform_lessons_US_abroad.pdf.

Gregorius, F. K. (1993). Disability insurance in the Netherlands. Insurance: Mathematics & Economics, 13:101–116.

Group Consultatif Actuariel Européen (2011). Use of age & disability as rating factors in insurance. Group Consultatif Actuariel Européen Position Paper. Available at: http://actuary.eu/documents/GC_Age_Disability_Underwriting_Paper_051211.pdf

Gruenberg, E. M. (1977). The failure of success. The Milbank Memorial Fund Quarterly. Health Society, 55(1):3–24.

Haberman, S. (1988). The central sickness rate: a mathematical investigation. In Transactions of the 23rd International Congress of Actuaries, volume 3, pages 83–99, Helsinki, Finland.

Haberman, S. (1996). Landmarks in the history of actuarial science (up to 1919). Dept. of Actuarial Science and Statistics, City University, London. Actuarial Research Paper No. 84. Available at: http://www.cass.city.ac.uk/__data/assets/pdf_file/0010/37198/84-ARC.pdf

Haberman, S., Butt, Z., and Rickayzen, B. (2004). Measuring process risk in Income Protection insurance. ASTIN Bulletin, 34(1):199–227.

Haberman, S. and Pitacco, E. (1999). Actuarial Models for Disability Insurance. Chapman & Hall/CRC, Boca Raton, USA.

Haberman, S. and Sibbett, T. A., editors (1995). History of Actuarial Science. Pickering & Chatto, London.

Hamilton-Jones, J. (1972). Actuarial aspects of long-term sickness insurance. Journal of the Institute of Actuaries, 98:17–67.

Hertzman, E. (1993). Summary of talk on "Work of the CMI PHI Sub-Committee". Insurance: Mathematics & Economics, 13:117–122.

Hoem, J. M. (1969). Markov chain models in life insurance. Blätter der deutschen Gesellschaft für Versicherungsmathematik, 9:91–107.

Hoem, J. M. (1972). Inhomogeneous semi-Markov processes, select actuarial tables, and duration-dependence in demography. In Greville, T. N. E., editor, Population Dynamics, pages 251–296. Academic Press, London.

Hoem, J. M. (1988). The versatility of the Markov chain as a tool in the mathematics of life insurance. In Transactions of the 23rd International Congress of Actuaries, volume R, pages 171–202, Helsinki.

IAA (2009). Note on Enterprise Risk Management for Capital and Solvency Purposes in the Insurance Industry. International Actuarial Association. Available online: http://www.actuaries.org/CTTEES_FINRISKS/Documents/Note_on_ERM.pdf

IAA (2011). Note on Enterprise Risk Management for Pensions. International Actuarial Association - Pensions and Employee Benefits Committee. Available online: http://www.actuaries.org/LIBRARY/Papers/Note_ERM_Pensions_EN.pdf

Jakubowski, E., Busse, R., Chambers, G. R. (1998). Health care systems in the EU. A comparative study. European Parliament, Public Health and Consumer Protection Series. Available at: http://www.europarl.europa.eu/workingpapers/saco/pdf/101_en.pdf

Janssen, J. (1966). Application des processus semi-markoviens à un problème d'invalidité. Bulletin de l'Association Royale des Actuaires Belges, 63:35–52.

Jean, J. (2004). Accident insurance. In Teugels, J. L. and Sundt, B., editors, Encyclopedia of Actuarial Science, volume 1, pages 1–4. John Wiley & Sons.

Jones, B. L. (1995). A stochastic population model for high demand CCRCs. Insurance: Mathematics & Economics, 16:69–77.

Jones, B. L. (1996). Transient results for a high demand CCRC model. Scandinavian Actuarial Journal, pages 65–182.

Jones, B. L. (1997a). Methods for the analysis of CCRC data. North American Actuarial Journal, 1(2):40–54.

Jones, B. L. (1997b). Stochastic models for continuing care retirement communities. North American Actuarial Journal, 1(1):50–73.

Jones, B. L. (2004). Long-term care insurance. In Teugels, J. L. and Sundt, B., editors, Encyclopedia of Actuarial Science, volume 2, pages 1037–1041. John Wiley & Sons.

Kimball, M., Phily, C., Folsom, A., Lagomarsino, G., Holtz, J. (2013). Leveraging Health Microinsurance to promote universal health coverage. Microinsurance Paper No. 23 - International Labour Organization. Available at: http://www.microinsurancefacility.org/publications/mp23

Kramer, M. (1980). The rising pandemic of mental disorders and associated chronic diseases and disabilities. Acta Psychiatrica Scandinavica, 62(S285):382–397.

Lafortune, G. and Balestat, G. (2004). Trends in severe disability among elderly people: assessing the evidence in 12 OECD countries and the future implications. OECD Health Working Papers, No. 26, OECD Publishing. Available at: http://www.oecd-ilibrary.org/social-issues-migration-health/trends-in-severe-disability-among-elderly-people_217072070078

Lawton, M. P. and Brody, E. M. (1969). Assessment of older people: self-maintaining and instrumental activities of daily living. Gerontologist, 9:179–186.

Le Roy, P. and Holtz, J. (2011). Third party payment mechanisms in Health Microinsurance. Microinsurance Paper No. 13 - International Labour Organization. Available at: http://www.microinsurancefacility.org/sites/default/files/mpaper13.pdf

Leatherman, S., Jones Christensen, L., and Holtz, J. (2010). Innovations and barriers in Health Microinsurance. Microinsurance Paper No. 6 - International Labour Organization. Available at: http://www.microinsurancefacility.org/sites/default/files/mpaper6.pdf

Leung, E. (2004). A multiple state model for pricing and reserving private long term care insurance contracts in Australia. Available at: http://www.actuaries.asn.au/Library/horizons_paper0604.pdf

Lurie, P. (2007). Actuarial methods in health insurance provisioning, pricing and forecasting. Institute of Actuaries of Australia, Sydney. Available at: http://www.actuaries.asn.au/Library/5.e_Conv07_Paper_Lurie_actuarial%20methods%20in%20health%20insurance%20pricing.pdf
Mackay, G. (1993). Permanent Health Insurance. Overviews and market conditions in the UK. Insurance: Mathematics & Economics, 13:123–130.
Manton, K. G. (1982). Changing concepts of morbidity and mortality in the elderly population. The Milbank Memorial Fund Quarterly. Health Society, 60(2):183–244.
Martin, J. and Elliot, D. (1992). Creating an overall measure of severity of disability for the Office of Population Censuses and Surveys disability survey. Journal of the Royal Statistical Society. Series A (Statistics in Society), 155(1):121–140.
Mattsson, P. (1977). Some reflections on different disability models. Scandinavian Actuarial Journal, pages 110–118.
McDowell, I. (2006). Measuring Health. A Guide to Rating Scales and Questionnaires. Oxford University Press.
Merlis, M. (2004). Long-Term Care financing: Models and issues. Working Paper. National Academy of Social Insurance. Washington, DC. Available at: http://www.nasi.org/usr_doc/Merlis_LongTerm_Care_Financing.pdf
Milbrodt, H. (2005). Aktuarielle Methoden der deutschen privaten Krankenversicherung. Schriftenreihe Angewandte Versicherungsmathematik, 34. Verlag Versicherungswirtschaft, Karlsruhe.
Murtaugh, C. M., Spillman, B. C., and Warshawsky, M. J. (2001). In sickness and in health: An annuity approach to financing Long-Term Care and retirement income. The Journal of Risk and Insurance, 68(2):225–254.
Newsom, M. and Fernandez, B. (2010). Private health insurance premiums and rate reviews. Congressional Research Service. Washington, DC. Available at: http://digitalcommons.ilr.cornell.edu/key_workplace/788
Nuttall, S. R., Blackwood, R. J. L., Bussell, B. M. H., Cliff, J. P., Cornall, M. J., Cowley, A., Gatenby, P. L., and Webber, J. M. (1994). Financing long-term care in Great Britain. Journal of the Institute of Actuaries, 121:1–68. Available at: http://www.actuaries.org.uk/research-and-resources/documents/financing-long-term-care-great-britain
OECD (2004a). Private Health Insurance in OECD Countries. Technical report, The OECD Health Project, OECD Publishing.
OECD (2004b). Proposal for a taxonomy of health insurance. Technical report, OECD Private Health Insurance Study. Available at: http://www.oecd.org/health
OECD (2013). Health at a Glance 2013. OECD INDICATORS. OECD Publishing. Available at: http://www.oecd.org/els/health-systems/Health-at-a-Glance-2013.pdf
O'Grady, F. T., editor (1988). Individual Health Insurance, Illinois, USA. The Society of Actuaries.
Olivieri, A. and Ferri, S. (2003). Mortality and disability risks in Long Term Care insurance. IAAHS Online Journal, (1). Available at: http://www.actuaries.org/IAAHS/OnlineJournal/2003-1/Mortality_and_Disability_Risks_in_Long_Term_Care_Insurance.pdf
Olivieri, A. and Pitacco, E. (2002a). Managing demographic risks in Long Term Care insurance. Rendiconti per gli Studi Economici Quantitativi, 2(2):15–37.
Olivieri, A. and Pitacco, E. (2002b). Premium systems for post-retirement sickness covers. Belgian Actuarial Bulletin, 2:15–25.
Olivieri, A. and Pitacco, E. (2009). Stochastic models for disability: Approximations and applications to sickness and personal accident insurance. The ICFAI University Journal of Risk & Insurance, V I(2):19–43. Available at SSRN: http://ssrn.com/abstract=1266098
Olivieri, A. and Pitacco, E. (2011). Introduction to Insurance Mathematics. Technical and Financial Features of Risk Transfers. Springer.
Orros, G. and Howell, J. (2008). Creating value through integrated ERM for healthcare insurers in Europe. In ERM Symposium, Chicago, Illinois. Available at: http://www.ermsymposium.org/2008/pdf/papers/Orros.pdf
Orros, G. and Smith, J. (2010). ERM for health insurance from an actuarial perspective. Sessional Research Paper, presented to the Institute and Faculty of Actuaries. Avail-

able at: http://www.actuaries.org.uk/research-and-resources/documents/erm-health-insurance-actuarial-perspective

Orros, G. and Webber, J. (1988). Medical expenses insurance - An actuarial review. Journal of the Institute of Actuaries, 115:169–269. Available at: http://www.actuaries.org.uk/research-and-resources/documents/medical-expenses-insurance-actuarial-review

Pitacco, E. (1992). Risk classification and experience rating in sickness insurance. In Transactions of the 24th International Congress of Actuaries, volume 3, pages 209–221, Montreal, Canada.

Pitacco, E. (1994). LTC insurance. From the multistate model to practical implementations. In Proceedings of the 25th International ASTIN Colloquium, pages 437–452, Cannes, France.

Pitacco, E. (1995). Actuarial models for pricing disability benefits: towards a unifying approach. Insurance: Mathematics & Economics, 16(1):39–62.

Pitacco, E. (1999). Multistate models for long-term care insurance and related indexing problems. Applied Stochastic Models in Business and Industry, 15:429–441.

Pitacco, E. (2004a). Disability insurance. In Teugels, J. L. and Sundt, B., editors, Encyclopedia of Actuarial Science, volume 1, pages 530–541. John Wiley & Sons.

Pitacco, E. (2004b). Disability insurance, numerical methods. In Teugels, J. L. and Sundt, B., editors, Encyclopedia of Actuarial Science, volume 1, pages 541–548. John Wiley & Sons.

Pitacco, E. (2012). Mortality of disabled people. Technical Report. Available at SSRN: http://ssrn.com/abstract=1992319, and at: http://www.actuaries.org/CTTEES_TFM/Documents/Mortality_Disabled.pdf

Pitacco, E. (2013). Biometric risk transfers in life annuities and pension products: a survey. CEPAR Working Paper 2013/25. Available at: http://www.cepar.edu.au/media/123434/25_biometric_risk_transfers_in_life_annuities_and_pension_products.pdf

Pitacco, E., Denuit, M., Haberman, S., and Olivieri, A. (2009). Modelling Longevity Dynamics for Pensions and Annuity Business. Oxford University Press.

Pott, J. and Holtz, J. (2013). Value-added services in Health Microinsurance. Microinsurance Paper No. 19 - International Labour Organization. Available at: http://www.microinsurancefacility.org/sites/default/files/mpaper19.pdf

Ramlau-Hansen, H. (1991). Distribution of surplus in life insurance. ASTIN Bulletin, 21:57–71.

Rickayzen, B. D. (2007). An analysis of disability-linked annuities. Faculty of Actuarial Science and Insurance, Cass Business School, City University, London. Actuarial Research Paper No. 180. Available at: http://www.cass.city.ac.uk/__data/assets/pdf_file/0018/37170/180ARP.pdf

Rickayzen, B. D. and Walsh, D. E. P. (2002). A multi-state model of disability for the United Kingdom: Implications for future need for Long-Term Care for the elderly. British Actuarial Journal, 8(36):341–393.

Ridsdale, B. (2012). Annuity underwriting in the United Kingdom. Note for the International Actuarial Association Mortality Working Group. Available at: http://www.actuaries.org/index.cfm?lang=EN&DSP=CTTEES_TFM&ACT=IB_underwriting

Riedel, O. (2006). Unisex tariffs in Health Insurance. The Geneva Papers, 31:233–244.

Rudolph, M. J. (2009). Enterprise Risk Management (ERM) practice as applied to health insurers, self-insured plans, and health finance professionals. Research Report. Society of Actuaries, Health Section. Available at SSRN: http://www.soa.org/files/pdf/research-erm-pract-health.pdf

Sanchez-Delgado, E., de Paz-Cobo, S., and Lopez-Zafra, J. M. (2009). Mortality rates of Spanish dependents: a joint correction approach. In Otamendi, J., Bargiela, A., Montes, J. L., and Pedrera, L. M. D., editors, Proceedings of the 23rd European Conference on Modelling and Simulation, Madrid, Spain. Available at: http://www.scs-europe.net/conf/ecms2009/ecms2009%20CD/ecms2009%20accepted%20papers/fe_0149_8234de37.pdf

Sand, R. and Riis, J. (1980). Some theoretical aspects and experience with a simplified method of premium rating in disability insurance. In Transactions of the 21st International Congress of Actuaries, volume 3, pages 251–263, Zürich-Lausanne, Switzerland.

Sanders, A. J. and Silby, N. F. (1988). Actuarial aspects of PHI in the UK. Journal of Staple Inn Actuarial Society, 31:1–57.

Seal, H. L. (1977). Studies in the history of probability and statistics. XXXV. Multiple decrements or competing risks. Biometrika, 64:429–439.

Segerer, G. (1993). The actuarial treatment of the disability risk in Germany, Austria and Switzerland. Insurance: Mathematics & Economics, 13:131–140.

Society of Actuaries Long-Term Care Insurance Valuation Methods Task Force (1995). Long-Term Care insurance valuation methods. Transactions of the Society of Actuaries, 47:599–773.

Szuch, W. (2004). Sickness insurance. In Teugels, J. L. and Sundt, B., editors, Encyclopedia of Actuarial Science, volume 3, pages 1558–1559. John Wiley & Sons.

Thomson, S. and Mossialos, E. (2009). Private health insurance in the European Union. LSE Health and Social Care - London School of Economics and Political Science. Available at: http://www.ec.europa.eu/social/BlobServlet?docId=4216&langId=en

Verbrugge, L. M. and Jette, A. M. (1994). The disablement process. Social Science & Medicine, 38(1):1–14.

Vercruysse, W., Dhaene, J., Denuit, M., Pitacco, E., Antonio, K. (2013). Premium indexing in lifelong health insurance. Far East Journal of Mathematical Sciences, Special Volume 2013, Part IV:365–384. Available at: http://www.econ.kuleuven.ac.be/tew/academic/actuawet/pdfs/13-FJMS-SVIV-365.pdf

Warshawsky, M. J. (2007). The life care annuity - A proposal for an insurance product innovation to simultaneously improve financing and benefit provision for long-term care and to insure the risk of outliving assets in retirement. Georgetown University - Long-Term Care Financing Project. Working Paper No. 2. Available at: http://ltc.georgetown.edu/forum/2warshawsky061107.pdf

Waters, H. R. (1984). An approach to the study of multiple state models. Journal of the Institute of Actuaries, 111:363–374.

Waters, H. R. (1989). Some aspects of the modelling of Permanent Health Insurance. Journal of the Institute of Actuaries, 116:611–624.

Waters, H. R. (1992). Non-reported claims in long term sickness insurance. In Transactions of the 24th International Congress of Actuaries, volume 2, pages 335–342, Montreal, Canada.

Wendt, C. (2009). Mapping European healthcare systems: a comparative analysis of financing, service provision and access to healthcare. Journal of European Social Policy, 19:432–445.

Werth, M. and Mannion, P. (1998). Critical Illness cover - A time for review. Available at: http://www.marclife.com/research/pdf/cicuk.pdf

Whynes, D. K. (1996). The provision and finance of long-term care in the United Kingdom. The Geneva Papers on Risk and Insurance, 21(79):271–283.

Winklevoss, H. E. and Powell, A. V. (1984). Continuing Care Retirement Communities. An empirical, financial, and legal analysis. R. D. Irwin, Homewood, Illinois.

Wismar, M., Palm, W., Figueras, J., Ernst, K., and van Ginneken, E., editors (2011). Cross-border Health Care in the Europe an Union. Observatory Studies Series, No. 22. European Observatory on Health Systems and Policies. Available at: http://www.euro.who.int/__data/assets/pdf_file/0004/135994/e94875.pdf

Yamamoto, D. H. (2013). Health care costs - From birth to death. Research Report. Society of Actuaries, Health Section. Available at SSRN: http://www.soa.org/files/research/projects/research-health-care-birth-death-report.pdf

Zayatz, T. (2005). Social Security disability insurance program worker experience. Actuarial Study 118, Social Security Administration, USA. Available at: http://www.ssa.gov/oact/NOTES/pdf_studies/study118.pdf

Zhou-Richter, T. and Gründl, H. (2011). Life Care annuities - Trick or treat for insurance companies? Technical Report. Available at SSRN: http://ssrn.com/abstract=1856994

Zweifel, P. (1996). Providing for long-term care: Insurance vs. trust saving. The Geneva Papers on Risk and Insurance, 21(79):284–292.

Index

E. Pitacco, *Health Insurance*, EAA Series,
DOI 10.1007/978-3-319-12235-9

W

Y

Z

MIX
Papier aus verantwortungsvollen Quellen
Paper from responsible sources
FSC® C105338

If you have any concerns about our products,
you can contact us on
ProductSafety@springernature.com

In case Publisher is established outside the EU,
the EU authorized representative is:
Springer Nature Customer Service Center GmbH
Europaplatz 3, 69115 Heidelberg, Germany

Printed by Libri Plureos GmbH
in Hamburg, Germany